SLIME MOLDS

A Handbook of Myxomycetes

FULL-COLOR EXPANDED EDITION

STEVEN L. STEPHENSON

LONE PINE

Lone Pine Publishing

Library and Archives Canada Cataloguing in Publication
Title: Slime molds : a handbook of myxomycetes / Steven L. Stephenson.
Names: Stephenson, Steven L., author.
Description: Includes bibliographical references and index.
Identifiers: Canadiana (print) 2023058442X | Canadiana (ebook) 20230584462 | ISBN 9781772131260 (softcover) | ISBN 9781772131277 (EPUB)
Subjects: LCSH: Myxomycetes.
Classification: LCC QK635.A1 S74 2024 | DDC 579.5/2—dc23

Illustration Credits: All drawings by Dr. Henry Stempen, except Dr. Dmytro Leontyev, 12, 19, 107, 109.

Photo Credits: Jorge Almeida, 156; Bob Armstrong, 47; Christoph Benisch, 45; Paul Bos, 159; Mark Bower, 72, 132; Randy Darrah, 18; Todd Elliott. 10, 72, 74, 124, 137; Kim Fleming, 18, 55, 58, 60, 62, 74, 81, 82, 94, 104, 136, 142, 146, 164; Jeffery Gruber, 59, 106; Emily Johnson, 162; Henri Koskinen, 18, 46, 59, 71, 76, 79, 114, 147, 148, 149, 151, 157, 168; Gary Laursen, 68, 122; Laurie Leonard, 58, 69, 95, 123, 166, 169; Dmitro Leontyev, 154, 164; Jonathan Mays, 58, 108; Raymond McNeil, 70, 133; Alain Michaud, 21, 22, 24, 56, 57, 59, 60, 64–69, 70–72, 77, 78, 86, 88, 90, 93, 97, 99, 100, 109–111, 119, 126, 127, 141, 144, 148–151, 153–156, 158–160, 162-165, 167, 169, 170; Alex Michaud, 156, 157; Juan Francisco Moreno, 74, 139, 150, 153, 158, 161, 166; Marius Muessler, 49, 57, 68, 74, 84, 92, 115, 163; Yuri Novozhilov, 57, 59, 63, 68, 70, 71, 89, 102, 103, 121, 125, 128, 155, 159, 161, 167, 171; Richard Orr, 62, 135; Alison Pollack, 59, 61, 80, 107, 117, 119, 154, 155, 161, 165, 165, 168; Carlos Rojas, 147, 151, 152; Bill Roody, 11; Anne-Marie Rantet Roux, 120; John Shadwick, 58, 134; Henry Stempen, 25; Steve Stephenson, 11, 18, 34, 35, 37, 38, 39, 40, 105; Joe Takano, 66, 69, 72, 101, 112, 130, 147, 152, 167; Steve Young, 16, 60, 64, 83, 117, 160, 171.

Every effort has been made to accurately credit the sources of photographs and illustrations. Any errors or omissions should be reported directly to the publisher for correction in future editions.

We acknowledge the financial support of the Government of Canada.
Nous reconnaissons l'appui financier du gouvernement du Canada.

Funded by the Government of Canada
Financé par le gouvernement du Canada | Canada

PC: 42-1

Contents

Preface

I hope this book provides anyone who uses it with a basis for appreciating the beauty and diversity of a truly fascinating but little-known group of organisms—the myxomycetes or slime molds. The book was developed with the intent of serving both as a field guide to the group for the general naturalist, amateur mycologist, or interested layperson and as a useful introductory text for anyone who might be engaged in more serious study of the cosmopolitan species of myxomycetes.

As a field guide to most of the common species of myxomycetes, this book is not a comprehensive treatment of the group, but virtually all the species of myxomycetes that one is most likely to encounter are included. For the most part, the descriptions, illustrations, and keys included in the book are based upon material collected in eastern North America, but since (as will be discussed later) most myxomycetes are cosmopolitan, with the same species likely to be found anywhere on earth where suitable conditions exist, the use of this book is not limited to any particular region of the world.

The book is divided into eight chapters. The first chapter provides an overview of myxomycetes, including a description of their life cycle, how they are classified, and what is involved in collecting and identifying these organisms. The second chapter explains the structure of myxomycetes; such information is an essential prerequisite for identifying these organisms. The third chapter describes various techniques used to study myxomycetes in the home or laboratory, including a method used to culture (literally "grow your own") myxomycetes from samples of tree bark or other types of organic debris. The fourth chapter provides information on the worldwide distribution of myxomycetes, both in terms of the major types of habitats in which they characteristically occur as well as some of the microhabitats in which at least some members of the group can be found. The fifth chapter discusses various aspects of myxomycete ecology, including the relationships that exist between myxomycetes and other kinds of living organisms. The sixth chapter describes the system of classification used for myxomycetes and dichotomous keys first to the different taxonomic groups of myxomycetes and then to selected species (a total of 58 primary species) within each of these groups. The seventh chapter consists of a set of descriptions of the 58 primary species included in the keys. Comments provided with each description allow other morphologically similar species to the identified. A pen and ink drawing and color image are provided for each species. The eighth chapter provides color images of an additional 74 species mentioned in the comments sections of the 58 primary species. A glossary of technical terms, a list of references cited throughout the text, and an index follow at the end of the book.

In preparing the book, I have made liberal use of the information found in a number of standard textbooks of mycology and several major taxonomic monographs of the myxomycetes. These include *A Monograph of the Mycetozoa* (Lister 1925), *The Mycetozoa of North America* (Hagelstein 1944), *Biology of the Myxomycetes* (Gray and Alexopoulos 1968), *The Myxomycetes* (Martin and Alexopoulos 1969), *Flora Neotropica Monograph No. 16 (Myxomycetes)* (Farr 1976), *Introductory Mycology* (Alexopoulos and Mims 1979), *The Genera of Myxomycetes* (Martin et al. 1983), *The Myxomycetes of Britain and Ireland: an Identification Handbook* (Ing 1999), *Les Myxomycètes* (Poulain et al. 2011), and *Secretive Slime Moulds: Myxomycetes of Australia* (Stephenson 2021).

I gratefully acknowledge the contributions that others have made to this book. The majority of the pen and ink drawings were prepared specifically for the book by the late Dr. Henry Stempen, but two (7-26B and 7-28B) were contributed by Dr. Dmytro Leontyev. The latter also produced the color sketches used for Figures 1-4 and 2-6. The color images of species of myxomycetes used herein were provided by numerous individuals. These included Jorge Almeida, Bob Armstrong, Christoph Benisch, Paul Bos, Mark Bower, Todd Elliott, Kim Fleming, Jeffrey Gruber, Emily Johnson, Henri Koskinen, Carlos Lado, Gary Laursen, Laurie Leonard, Dmytro Leontyev, Pu Liu, Jonathan Mays, Raymond McNeil, Alain Michard, Juan Francisco Moreno, Marius Muessler, Yuri Novozhilov, Richard Orr, Alison Pollack, Anne-Marie Rantet Poux, Carlos Rojas, John Shadwick, Joe Takano, and Steve Young. Special thanks are due to Alain Michard, who was able to meet many of my "image needs" as the book project developed.

I would also like to express my appreciation to Dr. Orson K. Miller, Jr., who first introduced me to the myxomycetes; Dr. Marie L. Farr, who played a significant role in expanding my knowledge of these organisms; Barbara C. Stephenson, without whose assistance, support, and encouragement this book would not have been possible; my former student Denise Binion, who typed the first draft of the manuscript of the original edition; and Heino Lepp, who reviewed the manuscript of the second edition and offered suggestions to improve it.

List of Primary Species

Species	Page #	Figure #
Lamproderma sauteri	111	6-33, 7-30A, 7-30B
Lamproderma scintillans	112	6-35, 7-31A, 7-31B
Leocarpus fragilis	114	6-52, 7-32A, 7-32B
Lepidoderma tigrinum	115	6-41, 7-33A, 7-33B
Licea biforis	117	6-21, 7-34, 7-36
Licea kleistobolus	117	6-24, 7-35, 7-36
Licea minima	119	6-22, 7-36, 7-37
Licea operculata	119	6-23, 7-36, 7-38
Lycogala epidendrum	120	1-1, 2-14, 7-39A, 7-39B
Metatrichia vesparia	121	6-15, 7-40A, 7-40B
Mucilago crustacea	122	6-39, 7-41A, 7-41B
Perichaena depressa	123	6-10, 7-42A, 7-42B
Physarella oblonga	124	6-56, 7-43A, 7-43B
Physarum albescens	125	6-53, 7-44A, 7-44B
Physarum album	126	6-58, 7-45A, 7-45B
Physarum bivalve	127	6-49, 7-46A, 7-46B
Physarum cinereum	128	6-51, 7-47A, 7-47B
Physarum globuliferum	130	6-59, 7-48A, 7-48B
Physarum polycephalum	132	6-57, 7-49A, 7-49B
Physarum virescens	133	6-50, 7-50A, 7-50B
Prototrichia metallica	134	6-11, 7-51A, 7-51B
Reticularia splendens	135	6-27, 7-52A, 7-52B
Stemonitis axifera	136	6-62, 7-53A, 7-53B
Stemonitis fusca	137	6-63, 7-54A, 7-54B
Stemonitopsis typhina	139	6-66, 7-55A, 7-55B
Trichia decipiens	141	6-20, 7-56A, 7-56B
Trichia favoginea	142	2-11, 6-12, 7-57A, 7-57B
Trichia varia	144	2-8, 6-13, 7-57, 7-58A, 7-58B
Tubifera ferruginosa	146	2-4, 6-26, 7-59A, 7-59B

List of Additional Species

Species	Page #	Figure #
Alwisia bombarda	147	8-1
Amaurochaete atra	147	8-2
Arcyria globosa	147	8-3
Arcyria incarnata	148	8-4
Arcyria obvelata	148	8-5
Arcyria pomiformis	148	8-6
Arcyria stipata	149	8-7
Badhamia affinis	149	8-8
Badhamia foliicola	149	2-12, 8-9
Badhamia melanospora	150	8-10
Badhamia nitens	150	2-17, 8-11
Badhamia panicea	150	8-12
Brefeldia maxima	151	6-61, 8-13
Calomyxa metallica	151	8-14
Ceratiomyxa morchella	151	8-15
Ceratiomyxa sphaerosperma	152	8-16
Collaria arcyrionema	152	8-17
Colloderma oculatum	152	8-18
Comatricha pulchella	153	8-19
Comatricha suksdorfii	153	8-20
Craterium aureum	153	8-21
Craterium leucocephalum	154	8-22
Craterium obovatum	154	8-23
Cribraria confusa	154	8-24
Cribraria minutissima	155	8-25
Cribraria rufa	155	6-4, 8-26
Diachea subsessilis	155	8-27
Diderma fallax	156	8-28
Diderma hemisphaericum	156	8-29
Diderma montanum	156	8-30
Diderma radiatum	157	8-31
Diderma spumarioides	157	8-32
Didymium anellus	157	8-33
Didymium clavus	158	8-34
Didymium difforme	158	8-35
Didymium dubium	158	6-38, 8-36

Species	Page #	Figure #
Didymium nigripes	159	6-37, 8-37
Didymium squamulosum	159	8-38
Didymium vaccinum	159	8-39
Elaeomyxa cerifera	160	8-40
Hemitrichia clavata	160	8-41
Lepidoderma chailletii	160	8-42
Lepidoderma trevelyanii	161	8-43
Licea succulenticola	161	8-44
Lycogala conicum	161	8-45
Lycogala flavofuscum	162	8-46
Macbrideola cornea	162	8-47
Meriderma carestiae	162	6-30, 8-48
Meriderma cribrarioides	163	6-31, 8-49
Metatrichia floriformis	163	8-50
Oligonema flavidum	163	8-51
Paradiacheopsis solitaria	164	8-52
Perichaena chrysosperma	164	8-53
Perichaena corticalis	164	8-54
Perichaena dictyonema	165	8-55
Physarum bitectum	165	8-56
Physarum compressum	165	8-57
Physarum crateriforme	166	8-58
Physarum decipiens	166	8-59
Physarum didermoides	166	8-60
Physarum echinosporum	167	8-61
Physarum melleum	167	8-62
Physarum pusillum	167	8-63
Physarum roseum	168	8-64
Physarum viride	168	8-65
Reticularia lycoperdon	168	2-15, 5-2, 8-66
Stemonaria longa	169	8-67
Stemonitis splendens	169	8-68
Trichia botrytis	169	8-69
Trichia contorta	170	8-70
Trichia scabra	170	8-71
Trichia subfusca	170	8-72
Trichia verrucosa	171	8-73
Willkommlangea reticulata	171	8-74

Figure 1-1. Fruiting bodies of *Lycogala epidendrum* on decaying wood. The earliest published description of a myxomycete (in 1654) appears to be this species (©Todd Elliott).

CHAPTER 1

Introduction to Myxomycetes

A decaying log or stump seems an unlikely place to find one of nature's most extraordinary creatures. If one searches carefully, however, during the summer and early autumn, especially after a period of rainy weather, almost any woodland will yield a number of the fruiting bodies of a truly remarkable group of organisms, the slime molds.

Slime molds, or myxomycetes, as biologists call them, may not have a particularly attractive name, but members of the group produce fruiting bodies that exhibit incredibly diverse forms and colors and are often objects of considerable beauty. Most myxomycetes are quite small, reaching no more than a millimeter or two in height. Although large enough to be seen with the naked eye (Figure 1-1), the fruiting bodies of myxomycetes can best be observed with a hand lens or microscope. Only then can their intricate nature be fully appreciated. Fruiting bodies may take the shape of tiny goblets, globes, plumes, or other shapes more difficult to characterize. Some occur in tightly packed clusters, while others are scattered or even solitary. Many of the more intricate forms have a spore case held aloft on a delicate stalk, but others are attached directly to the substrate by their bases. Whatever form the fruiting body takes, its primary function is to produce the spores by which the myxomycete is propagated.

Myxomycetes have long intrigued and perplexed biologists because they possess characteristics of both animals and fungi. The fruiting bodies and spores they produce resemble those of many fungi, but some of their other attributes, including the capability for locomotion, are normally associated with animals.

For most of its life, a myxomycete exists as a thin, free-living mass of protoplasm. Sometimes this mass is several centimeters across and, as the name slime mold suggests, viscous and slimy to the touch. The mass of protoplasm, which is called a **plasmodium** (plural: **plasmodia**), can change form and creep slowly over the substrate upon which it occurs, much like a giant amoeba. As it moves, it feeds by engulfing bacteria and tiny bits of organic matter, another animal-like feature (Figure 1-2).

Figure 1-2. Myxomycete plasmodium moving across a decaying log (©Steve Stephenson).

Myxomycete plasmodia occur in cool, moist, shady places such as within crevices of decaying wood, beneath the partially decayed bark of logs and stumps, and in leaf litter. Consequently, they are not seen as frequently as are the fruiting bodies. Nonetheless, one can sometimes find plasmodia by stripping the bark from a decaying log or by carefully searching the moist underside of a piece of wood that has been lying on the forest floor. Plasmodia may be colorless or, as is more often the case, strikingly colored yellow, orange, or red.

After a period of feeding and growth, the plasmodium moves out of its normal habitat and into a drier, more exposed location. Here it gives rise to one or more fruiting bodies (Figure 1-3). This remarkable transformation from an animal-like to a fungus-like form seems more like something from science fiction than like science fact. Each fruiting body contains numerous spores, which are dispersed by the wind and, under suitable conditions, germinate, the first stage in development of a new plasmodium. Most people overlook myxomycetes because of their small size and because of where they occur. As anyone who has taken the time to become familiar with the group will undoubtedly attest, however, myxomycetes are among the most fascinating inhabitants of the woodlands of the world.

What Are Myxomycetes?

Figure 1-3. Plasmodium of *Fuligo septica* developing into the fruiting body (an aethalium) characteristic of this species (©Bill Roody).

Myxomycetes are a small, relatively homogeneous group of eukaryotic organisms. Only about 1200 species are known (Lado 2005–2022). The name given to the group is derived from the Greek words *myxa* (slime) and *mykes* (fungi) and was first used by German botanist Heinrich Link (1833). Link regarded the myxomycetes as fungi, a view that has not always been shared by other biologists. Anton de Bary, who is generally given credit for the first serious studies of the group, considered the myxomycetes to be more closely related to the protozoans and proposed the name Mycetozoa, which was derived from the Greek words *zoon* (animal) and *mykes* (fungus). Since de Bary, a number of monographers of the group, including Lister (1894), Hagelstein (1944), and Olive (1975), have perpetuated the name Mycetozoa. On the other hand, Macbride (1899), Macbride and Martin (1934), Martin (1949), Alexopoulos (1963), Martin and Alexopoulos (1969), Farr (1976), Nannenga-Bremekamp (1991), Neubert et al. (1993, 1995, 2000), Ing (1999), Stephenson (2003), Poulain et al. (2011), Yamamoto (2021), and Stephenson (2021) have followed Link's use of the term myxomycetes, although these organisms are no longer classified with the fungi. Whatever the relationship of myxomycetes to other living organisms, the study of these organisms has been, and still is, carried out by botanists, particularly mycologists (Martin et al. 1983).

Life Cycle

The life cycle of a typical myxomycete involves two distinct stages—the plasmodium and one or more fruiting bodies that develop from it. Actually, the complete life cycle is somewhat more complicated and includes an additional unicellular stage that is too small to be observed in nature. The sequence of events in the myxomycete life cycle, which is outlined in Figure 1-4, begins with a microscopic spore that was formed within and then released from one of the fruiting bodies characteristically produced by this group of organisms. Under suitable conditions, the spore germinates to produce one to four haploid, unwalled protoplasts (or cells). These are liberated through a small pore that forms in the spore wall or as a result of the spore cracking open.

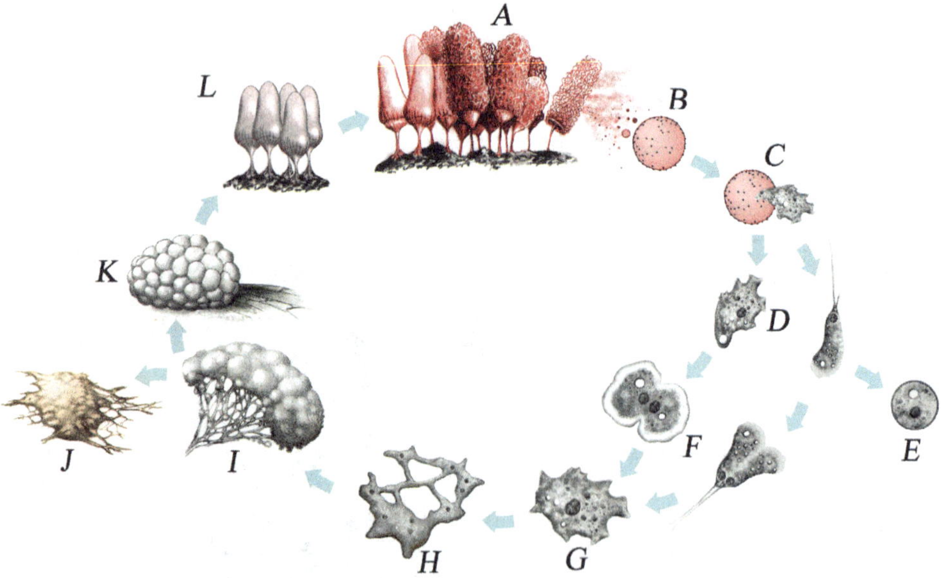

Figure 1-4. Life cycle of a typical myxomycete. (A) Group of fruiting bodies. (B) Spore. (C) A protoplast emerges from the spore. (D) The protoplast can take the form of an amoeba (left) or a flagellated cell (right) during the first trophic stage (the term amoeboflagellate applies to both forms). (E) Under dry conditions or in the absence of food, an amoeboflagellate can form a microcyst, or resting stage. (F) Comparable amoeboflagellates fuse to form a zygote. (G) Zygote. (H) The nucleus of the zygote divides by mitosis, and each subsequent nucleus also divides without being followed by cytokinesis, thus producing a single large cell, the plasmodium. (I) The plasmodium, which represents the second trophic stage in the life cycle. (J) Under adverse conditions, the plasmodium can form the second type of resting stage found in myxomycetes, the sclerotium. (K-L) Fruiting bodies develop from the plasmodium. During fruiting body formation, spores are produced (© Dmytro Leontyev).

Some protoplasts become flagellated soon after being released while others remain amoeboid. The latter may sometimes develop flagella (singular: flagellum) after a short time or, in some cases, simply remain amoeboid. Flagellated cells are called **swarm cells**, while the nonflagellated cells are called **myxamoebae** (singular: **myxamoeba**). Myxamoebae and swarm cells are interconvertible, and the particular form in which a given cell exists apparently depends to a large extent upon the availability of free water in its immediate environment. Because swarm cells and myxamoebae are interconvertible, the general term **amoeboflagellate cell** is sometimes used to refer to either type of cell. In the presence of free water, the flagellated form tends to predominate, while under drier conditions most cells exist in the nonflagellated amoeboid form.

Both myxamoebae and swarm cells divide by binary fission, with the flagella of the latter being retracted before each division. Under conditions unfavorable for continued growth or metabolism, myxamoebae undergo a reversible transformation to dormant structures called **microcysts**. Microcysts can remain viable for long periods of time and are probably very important in the continued survival of myxomycetes in some habitats such as deserts.

Ultimately, two compatible myxamoebae or swarm cells fuse to form a **zygote**. This process involves both the fusion of the protoplasm of the two cells (**plasmogamy**) as well as fusion of their nuclei (**karyogamy**). The resulting diploid zygote is at first either amoeboid or flagellated, depending upon the nature of the two cells that were involved in its formation. If flagellated, the zygote soon becomes amoeboid. The zygote then feeds and grows, ultimately to give rise to the plasmodium.

Growth of the zygote and the plasmodium that develops from it is accompanied by a series of synchronous nuclear divisions; as a result, the plasmodium becomes a multinucleate structure. Since individual cells are never delimited, however, the plasmodium remains acellular, that is, it is essentially a single giant cell. At maturity, a relatively small plasmodium contains several hundred nuclei; the number present in a large plasmodium is almost incomprehensible!

Under adverse conditions, such as drying out of the immediate environment or low temperatures, a plasmodium may convert into a hardened, resistant structure called a **sclerotium** (plural: **sclerotia**). Sclerotia are irregular masses of numerous small, cell-like units called **macrocysts**, which are capable of reforming the plasmodium under favorable conditions. During the winter months and other periods when myxomycetes are normally inactive, it is often possible to locate sclerotia by carefully removing the loose bark from decaying logs and stumps.

Eventually, a mature plasmodium undergoes the remarkable transformation that results in the production of one or more fruiting bodies. It is still not known what factors are involved in triggering the fruiting process (or **sporulation**), but laboratory studies indicate that exhaustion of the available food supply and changes in moisture, temperature, and pH can induce fruiting in some species. For species with pigmented plasmodia, exposure to light is often necessary before fruiting will occur. Once fruiting is initiated, the process appears to be irreversible and the entire plasmodium is converted into fruiting bodies typical of the particular species. The formation of each fruiting body is accompanied by cleavage of much of its protoplasm into uninucleate portions that mature into spores. Meiosis (reduction division) takes place in the young spores, which then become haploid. Typically, spores are uninucleate, but this is not always the case. When the fruiting body is fully mature, the spores are ready to be released and dispersed.

Until quite recently, it was thought that the life cycle of all myxomycetes followed the basic pattern outlined above, with compatible haploid uninucleate amoeboflagellate cells acting as gametes and fusing to produce a diploid zygote, which in turn gave rise to a plasmodium. The haploid condition was then restored as a result of meiosis occurring in the formation of spores in the fruiting bodies derived from the plasmodium. It is now known, however, that at least some species of myxomycetes have an alternate reproductive system in which there is no fusion of amoeboflagellate cells and in which meiosis does not occur during the formation of spores. Consequently, all the stages in the life cycle are diploid. This second type of life cycle is referred to as **apomictic**. The term is derived from the Greek words *apo* (away from) and *mixis* (mingling) and refers to the fact that sexual reproduction involving meiosis and the fusion of gametes does not occur. Both types of life cycles have been found to exist in some species of myxomycetes (Clark and Collins 1976), and it is even possible for a conversion from one to another to occur (Collins 1980). For at least some species of myxomycetes, the apomictic life cycle is likely to be the predominant type found in nature (Clark and Haskins 2010).

Other Slime Molds

The term slime mold has been applied to two rather different groups of organisms. The first of these is the group being considered in this book, and the second is a group of organisms known as **cellular slime molds** or **dictyostelids**. Myxomycetes are sometimes referred to as "true" slime molds, acellular slime molds, myxogastrids, or plasmodial slime molds to distinguish them from this other group. Cellular slime molds are single-celled,

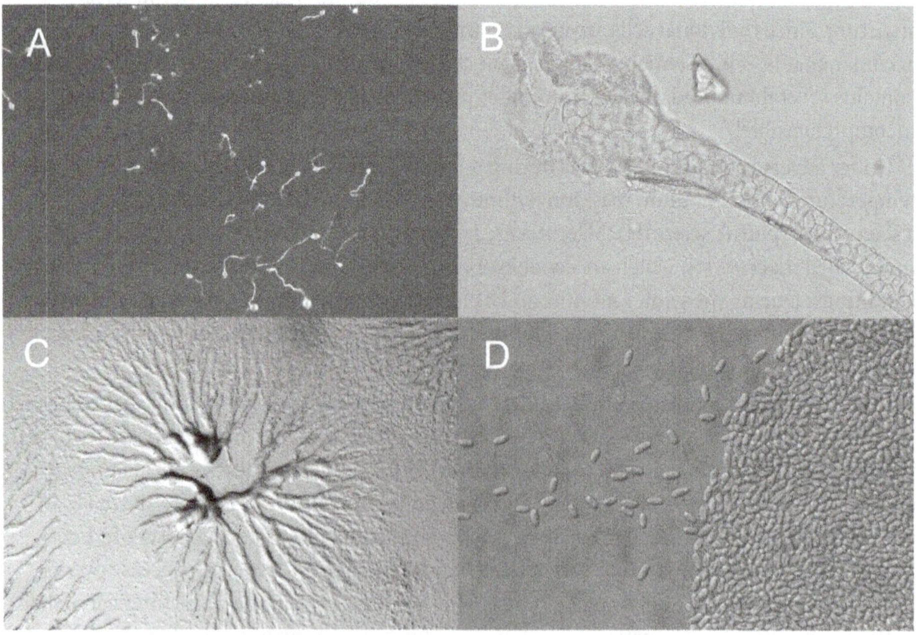

Figure 1-5. Major features in the life cycle of a cellular slime mold. (A) Mature sorocarps. (B) Stalk (or **sorophore**). (C) Spores. (D) Aggregation center (©Pu Liu)

soil-inhabiting predators of bacteria. These organisms represent a normal component of the microflora of soils and apparently play a role in maintaining the natural balance that exists between bacteria and other microorganisms in the soil microhabitat.

For most of their life cycle, cellular slime molds exist as independent, amoeboid cells (myxamoebae) that feed upon bacteria, grow, and multiply by binary fission. Except that they are never flagellated, these cells are fairly similar to the myxamoebae of myxomycetes. When the available food supply within a given microsite becomes depleted, numerous myxamoebae aggregate to form a multicellular structure (the **pseudoplasmodium**) in which each cell maintains its individual integrity. The pseudoplasmodium then produces one or more fruiting bodies (called **sorocarps** in this group) bearing spores. Under favorable conditions, the spores germinate to give rise to new myxamoebae, and the life cycle begins anew (Figure 1-5).

The cellular slime molds are a much smaller group than the myxomycetes, and only about 170 species are known to science (Raper 1984; Shrikh et al. 2017). Cellular slime molds are most abundant in the surface humus layers of forest soils, where bacterial populations are the highest and microenvironmental conditions appear to be most suitable for their growth and development. Quite commonly, six or more species can be found occurring together in the same sample of forest soil.

Biological Research with Myxomycetes

Because the myxomycete plasmodium essentially represents a single giant cell, it has often been used in laboratory studies of various aspects of cellular physiology and biochemistry. For example, the fact that the many nuclei of a plasmodium divide in natural synchrony every 8 to 10 hours provides unique opportunities to examine the events associated with the cell cycle, while the vigorous protoplasmic streaming exhibited by some types of plasmodia represents an ideal situation for studying cellular motility.

Moreover, the two major transformations that myxomycetes undergo in their life cycle—from microscopic, uninucleate amoeboflagellate cells to macroscopic, multinucleate plasmodia and then from plasmodia to fruiting bodies—make myxomycetes especially suitable as experimental organisms for studies of cell growth and differentiation. Since the apparently irreversible biochemical changes that occur when a plasmodium differentiates into fruiting bodies are at least suggestive of the changes that take place when a normal cell becomes cancerous, myxomycetes have also been used in cell differentiation studies connected with cancer research (Weaver 1976; Sun et al. 2017). More comprehensive albeit technical discussions of the use of myxomycetes as experimental organisms are provided by Aldrich and Daniel (1982), Dove et al. (1986), Kryvomaz et al. (2021), and Walker et al. (2021).

Figure 2-1. Plasmodium in nature (© Steve Young)

CHAPTER 2

Structural Features of Myxomycetes

Identification of myxomycetes is based upon such features as the gross structure of the fruiting body, the presence or absence of deposits of lime (calcium carbonate), and spore color in mass. Although some important diagnostic features can be observed with a hand lens or magnifying glass, many others are evident only when material mounted on a slide is examined with a microscope. Practically all these features are readily discernible with the magnifications possible using a standard laboratory microscope of the type available in most schools, although having an instrument equipped with an ocular micrometer and an oil immersion objective is a prerequisite for the accurate determination of some features.

Since identification is based almost exclusively upon features of fully mature fruiting bodies, both the major types of fruiting bodies produced by myxomycetes and the various components that make up a fruiting body will be described in some detail in this chapter. Although myxomycetes cannot be positively identified while still in their plasmodial stage, all the structures upon which identification is based originate in the plasmodium. Consequently, it would seem worthwhile to begin any discussion of the structural features of myxomycetes with a consideration of plasmodia.

The Plasmodium

The myxomycete plasmodium is an acellular, multinucleate mass of protoplasm bounded by a cell membrane and, for some types, enclosed by a slime sheath. Some of the material in the slime sheath is shed behind an advancing plasmodium in the form of readily visible "slime tracks" on the substrate. Close scrutiny of the inner surface of a piece of loose bark removed from a decaying log or a dead leaf taken from within the layer of leaf litter that has accumulated in a moist depression on the forest floor often reveals this indirect evidence of the presence of myxomycete plasmodia.

Plasmodia vary in color, size, and structure. Some plasmodia are hyaline or white but others are brightly colored (Figure 2-1). Yellow, orange, and red are some of the colors one is likely to encounter in nature. Color appears to be reasonably constant for a given species but is affected to some extent by factors such as temperature, the pH of the substrate, and various substances taken up by the plasmodium. For example, the ingestion of living algal cells may sometimes impart a green color to a plasmodium (Lazo 1961).

Four general types of plasmodia are recognized (Alexopoulos 1960). Virtually all the plasmodia that one is likely to encounter in nature belong to the type known as a **phaneroplasmodium**. In fact, the name is derived from the Greek word *phaneros* (visible). A typical phaneroplasmodium consists of a network of vein-like strands of protoplasm. The protoplasm is rather granular and exhibits rapid, reversible streaming within the network of veins. Migrating plasmodia of this type are often more or less fan-shaped, with the broad side of the "fan" representing the advancing anterior margin of the plasmodium. Phaneroplasmodia are particularly characteristic of members of the traditional order Physarales.

The second type of plasmodium, the **aphanoplasmodium**, is rarely observed in nature. This fact is suggested by its name, which is derived from the Greek word *aphanes* (invisible). Although consisting of a network of vein-like strands of protoplasm, the protoplasm is not very granular and the veins are very fine and lack a slime sheath. This combination of features renders the aphanoplasmodium rather inconspicuous. Aphanoplasmodia are characteristic of members of the traditional order Stemonitales.

The third and most primitive type of plasmodium is the **protoplasmodium**, with a name derived from the Greek word *protos* (first). A protoplasmodium remains microscopic throughout its entire existence and gives rise to only a single, rather minute fruiting body. This type of plasmodium forms no veins and the protoplasm exhibits a very slow irregular streaming instead of the rapid, rhythmical, reversible streaming of the other types. Members of the traditional order Echinosteliales and many of the smaller members of the traditional order Liceales produce protoplasmodia.

The fourth type of plasmodium, which appears to possess features of both phaneroplasmodia and aphanoplasmodia, is produced by members of the traditional order Trichiales (Alexopoulos 1960; McManus 1962). Referred to as a **trichiaceous plasmodium**, it tends to be rather inconspicuous in nature and is not capable of migrating very far within a particular substrate. Because few members of this order have been grown in culture, less is known about this type of plasmodium than about the other types.

Types of Fruiting Bodies

Myxomycetes produce four types of fruiting bodies or **fructifications**, as they are sometimes called. The first of these, the **sporangium** (plural: **sporangia**) is by far the most common (Figure 2-2). Sporangia are relatively small fruiting bodies of fairly uniform size and shape for a particular species. They are formed when a plasmodium breaks up into a number of small portions, each of which develops into a single sporangium. Because they are products of the same plasmodium, sporangia are usually clustered on a given substrate. Depending on the species, sporangia may be sessile or stalked, and a wide range of different shapes and colors are represented by the various myxomycetes that produce this type of fruiting body. It should be noted that Robbrecht (1974) proposed the term **sporocyst**—literally a spore box with an acellular wall—for the type of fruiting body traditionally referred to as a sporangium, and others, such as Buyck (1988), have followed his example. Another term that has been used for the fruiting body of a myxomycete is **myxocarp** (Poulain et al. 2011).

The second type, the **aethalium** (plural: **aethalia**), is a comparatively large fruiting body that is usually hemispherical or cushion-shaped, always sessile, and often rather variable in size (Figure 2-3). It is derived from all or a major portion of a plasmodium. Aethalia are thought by some to represent masses of completely fused sporangia.

Figure 2-2. Developing sporangia of *Cribraria argillacea* (©Steve Stephenson).

Figure 2-3. Aethalium of *Fuligo septica* (©Henri Koskinen).

Figure 2-4. Pseudoaethalium of *Tubifera ferruginosa* in which the sporangium-like units are clearly apparent (©Kim Fleming).

Figure 2-5. Unusually large plasmodiocarp of *Hemitrichia serpula* (©Randy Darrah).

The third type of fruiting body, the **pseudoaethalium** (plural: **pseudoaethalia**), is a composite structure derived from several to many sporangium-like units crowded together. The individual sporangium-like units are usually more or less distinguishable, at least in the early stages of development (Figure 2-4). A pseudoaethalium (literally, false aethalium) superficially resembles an aethalium, hence the name. Most pseudoaethalia are sessile, but a few examples are stalked.

The fourth type of fruiting body, the **plasmodiocarp**, is somewhat more difficult to characterize. A plasmodiocarp is a fruiting body that retains the shape of one or more of the main veins of the plasmodium from which it was derived (Figure 2-5). Some plasmodiocarps are extensive, netlike structures, but most are elongated, curved or straight, and simple or branched sporangium-like units. In fact, it is sometimes rather difficult to know where to draw the line between short plasmodiocarps and sessile sporangia, and the two types of fruiting bodies actually intergrade in some species of myxomycetes. Except for a few rather special examples, plasmodiocarps are sessile.

Components of a Fruiting Body

A fruiting body typically consists of six major components: sporotheca, stalk, hypothallus, peridium, columella (or sometimes a pseudocolumella), capillitium, and spores (or spore mass, when considered collectively) (Figure 2-6). Not all these components are present in all types of fruiting bodies, and some fruiting bodies have an additional component, either a **pseudocolumella** or a **pseudocapillitium** (Figure 2-7).

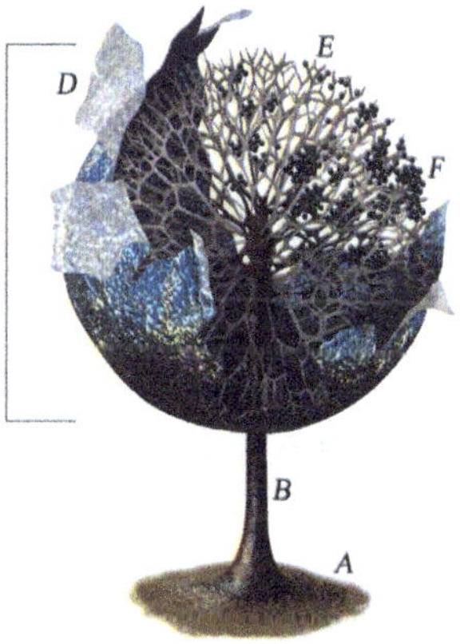

Figure 2-6. Structural features of the fruiting body of a myxomycete. (A) Hypothallus. (B) Stalk (sometimes this can extend into the spore mass as a columella). (C) Sporotheca. (D) Peridium. (E) Capillitium. (F) Spores (©Dmytro Leontyev).

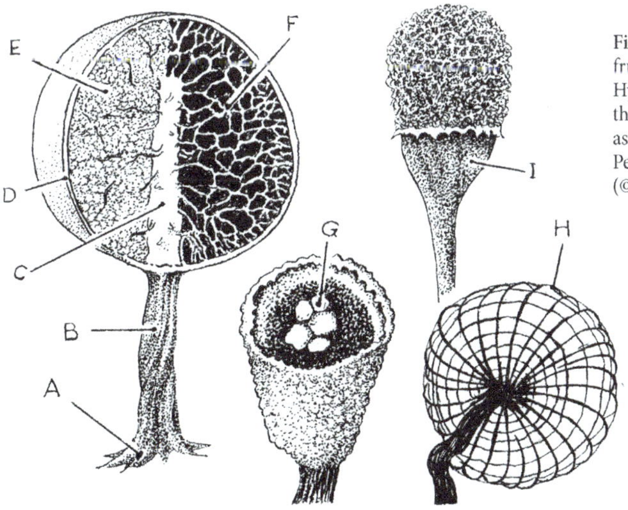

Figure 2-7. Structural features as expressed in the fruiting bodies (sporangia in the examples shown here) of different species of myxomycetes. Drawings are not to scale. (A) Hypothallus. (B) Stalk. (C) Columella. (D) Peridium. (E) Spores. (F) Capillitium (spores removed). (G) Pseudocolumella. (H) Peridial (surface) net. (I) Calyculus (©Henry Stempen).

Sporotheca

The actual spore-containing portion of a sporangium, as opposed to the entire structure including the stalk if the latter structure is present, is referred to as the **sporotheca** (Lado and Pando 1997) or, as noted above, the sporocyst (Poulain et al. 2011). The use of the term sporotheca is relatively recent, and earlier treatments of the myxomycetes, including the first edition of this book, considered "sporangium" to represent the entire fruiting body (i.e., sporotheca and stalk). This often made the figure given for height in the descriptions of stalked fruiting bodies confusing—did it apply only the upper portion containing the spore mass or to the entire structure. This distinction is not necessary for sessile fruiting bodies.

Stalk

In certain fruiting bodies the spore mass is elevated some distance above the substrate by a **stalk** (sometimes also called a **stipe** or **sorophore**). Stalks display considerable variation in length, thickness, color, and texture. Moreover, they may be hollow or filled with one of several types of material (for example, round, spore-like structures or granular debris), translucent or opaque, and covered with lime or limeless. Any of these can represent an important feature that must be considered in the identification of some myxomycetes.

Hypothallus

The **hypothallus** is a thin layer deposited by the plasmodium at the time of fruiting. It has many different expressions, ranging from membranous to relatively thick, delicate to tough, and more or less transparent to brightly colored. The hypothallus may surround the base of a single fruiting body or form a continuous layer connecting the bases of a number of fruiting bodies. In some myxomycetes it may not be evident at all.

Peridium

The **peridium** (plural: **peridi**a) is a covering that encloses the spore mass (plus any other associated structures) of a sporotheca. It varies in structure from delicate, thin, and membranous to relatively tough and thick. In some myxomycetes, the peridium is not evident in mature fruiting bodies, while in others only remnants of the peridium persist at maturity. If just the basal portion of the peridium persists as a cup-shaped structure around the bottom of the spore mass, the fruiting body is said to have a **calyculus**.

The manner in which the peridium breaks apart to reveal the spore mass is often an important diagnostic feature. Some fruiting bodies open by means of a preformed lid, while in others the peridium splits irregularly or along definite lines of dehiscence. The rather thick covering that surrounds the spore mass in an aethalium is usually referred to as a **cortex**.

Columella and Pseudocolumella

In some fruiting bodies, a portion of the space within the spore mass is occupied by a structure referred to as a **columella**. In stalked fruiting bodies, the columella appears to represent an extension of the stalk that simply continues upward into the spore mass, although it may or may not resemble the stalk in structure, shape, texture, and color.

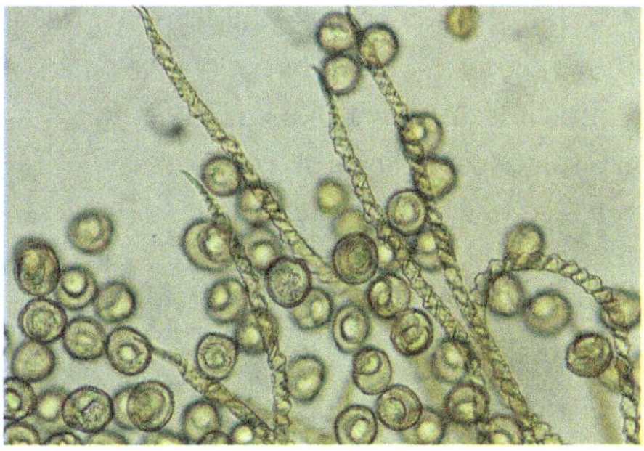

Figure 2-8. Capillitium and spores of *Trichia varia*. The capillitium is marked with loosely and irregularly wound spiral bands (630X) (©Alain Michaud).

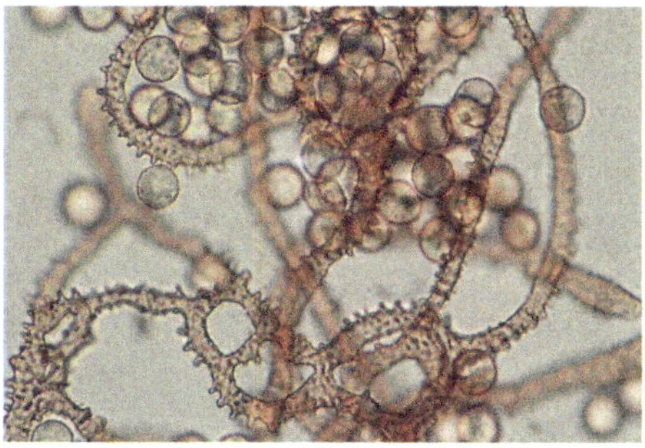

Figure 2-9. Capillitium and spores of *Arcyria denudata*. The capillitium is marked with spirally arranged cogs and half-rings (630X) (©Alain Michaud).

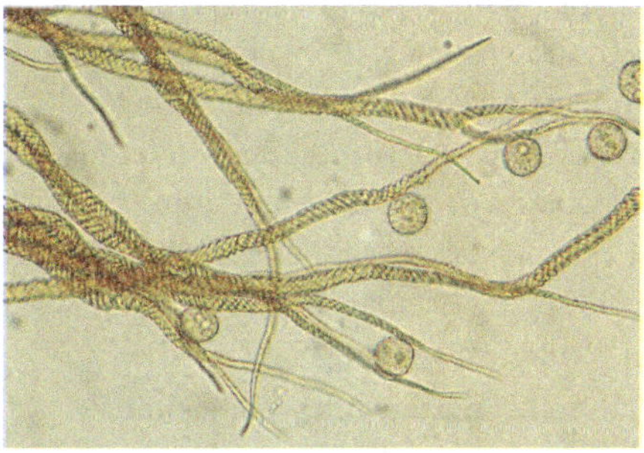

Figure 2-10. Capillitium of *Trichia crateriformis*. The individual capillitial elements (elaters) have elongated tips (630X) (©Alain Michaud).

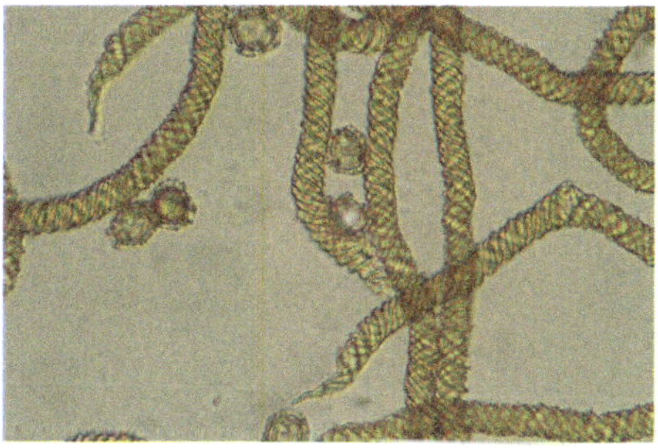

Figure 2-11. Capillitium of *Trichia favoginea*. In this species the individual capillitial elements (elaters) have short tips (630X) (©Alain Michaud).

In sessile fruiting bodies, the columella ranges from little more than a thickening of the inner side of the peridium where it contacts the substrate to a conspicuous dome-shaped structure. A **pseudocolumella** (literally, a false columella) is a structure that resembles a columella but is not attached in any way at the base of the spore mass. The usual expression is for the pseudocolumella, which is found only in the traditional order Physarales, to consist of a spherical or irregular lime mass centrally located within the spore mass. The lime mass may consist of a single discrete body or may be represented by a loose aggregation of smaller units. In many instances, portions of the capillitium (described below) are connected to the columella or pseudocolumella.

Capillitium and Pseudocapillitium

The capillitium (plural: capillitia) is a system of threadlike elements found within the spore mass of the fruiting body in many different species of myxomycetes (Figure 2-8). It may take the form of a dense network of interconnected elements (Figure 2-9) or consist of a series of free elements referred to as **elaters** (Figures 2-10 and 2-11). Portions of the capillitium may be solid or hollow, free or interconnected, smooth or sculptured, and contain lime or be limeless. A capillitium containing lime is present only in some members of the traditional

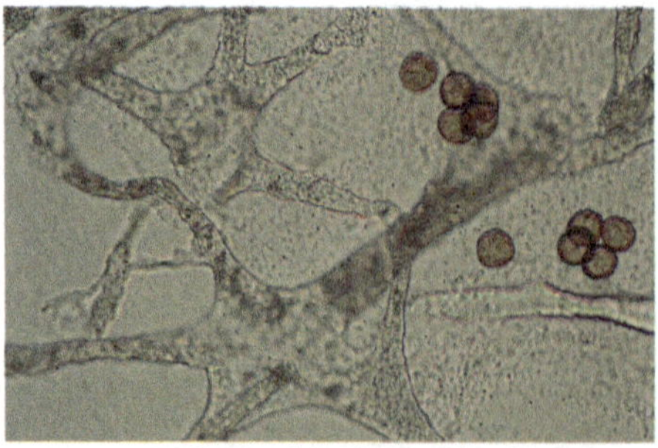

Figure 2-12. Badhamioid capillitium of *Badhamia foliicola* consisting entirely of lime (630X) (©Alain Michaud).

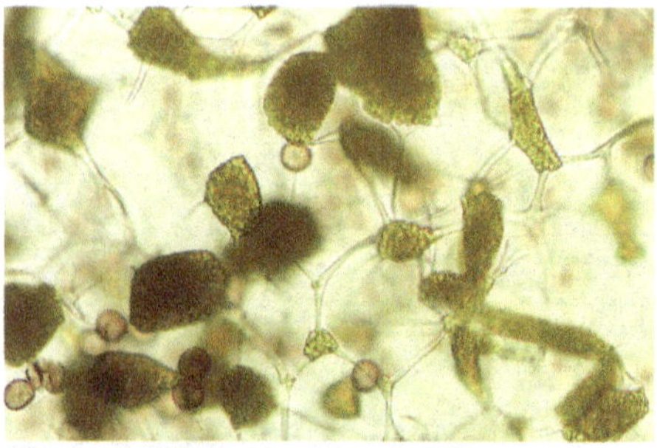

Figure 2-13. Physaroid capillitium of *Physarum citrinum* consisting of rounded lime nodes and delicate limeless connecting threads (630X) (©Alain Michaud).

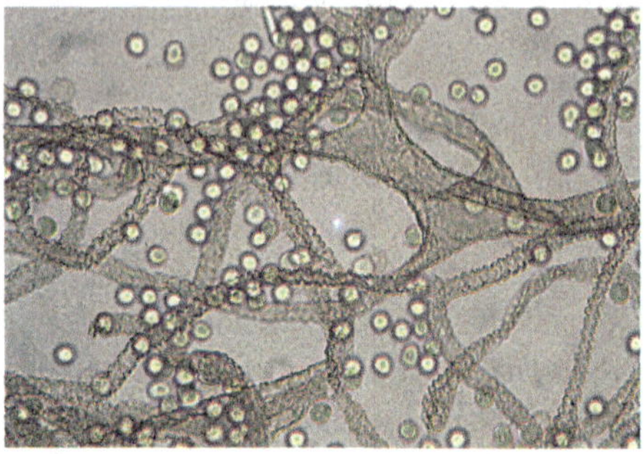

Figure 2-14. Pseudocapillitium of *Lycogala epidendrum* (400X) (©Alain Michaud).

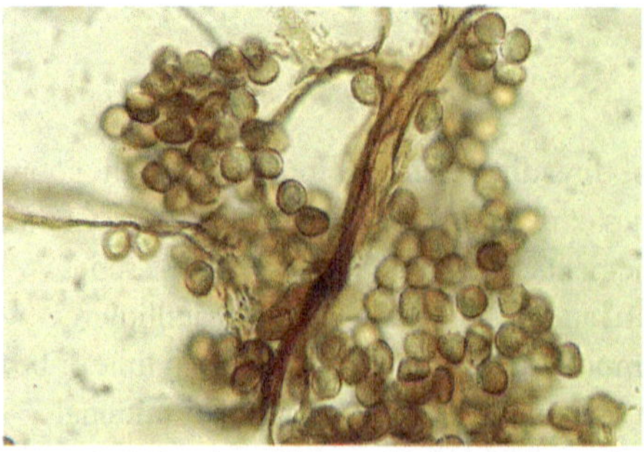

Figure 2-15. Pseudocapillitium of *Reticularia lycoperdon* (400X) (©Alain Michaud).

order Physarales and may be entirely limy (Figure 2-12) or consist of a system of limeless, hyaline tubules connecting conspicuous deposits of lime usually referred to as **lime nodes** (Figure 2-13). The term **badhamioid** is used to denote the former condition, while the term **physaroid** is used to denote the latter.

Although the capillitial elements are intermingled with the spores, they are in no way connected to them. In fact, the capillitium originates as a completely separate system in the developing fruiting body. It is generally thought to be formed from material deposited in tubular vacuoles that originate either within the protoplasm of a developing fruiting body or as invaginations of the cell membrane surrounding the mass of protoplasm that is to become the fruiting body.

In some myxomycetes, the capillitial elements are highly elastic and/or hygroscopic. In those fruiting bodies with a highly elastic capillitium the latter expands as the peridium breaks up, which tends to help disseminate the spores. If the capillitial elements are hygroscopic, they twist and turn in such a way as to expel spores forcibly from the spore mass.

A pseudocapillitium occurs only in some aethalia and pseudoaethalia, particularly those produced by certain members of the traditional order Liceales. Unlike the capillitium, where the individual elements are more or less uniform with respect to both overall structure and diameter, the elements making up the pseudocapillitium are distinctly irregular and vary in width or diameter (Figure 2-14). The pseudocapillitium may take the form of threadlike elements, bristles, membranes, or perforated plates (Figure 2-15). These are thought to originate from collapsed strands of the plasmodium (that is, essentially residual material) or portions of the sporangium-like units that have persisted in mature fructifications and not from a directed biosynthesis as is the case for a true capillitium.

However, the classic definition of what is considered to be a pseudocapillitium (*sensu* Lister 1925; Nannenga-Bremekamp 1991; Neubert et al. 1993; Poulain et al. 2011) was challenged by Leontyev et al. (2014a), who carried out a detailed investigation of the "bristles" that arise from the sporangium-like units of *Alwesia bombarda*. They pointed out that the basic structure of these bristles was not unlike what has been considered as "true" capillitium in some other species of myxomycetes. Moreover, they went on to suggest that the tubular elements of what has long been regarded as a pseudocapillitium in species of *Lycogala* might be better regarded as a capillitium. Clearly, this is one aspect of the morphology of myxomycetes that warrants additional study.

Spores

The spores of virtually all myxomycetes are spherical, and those of most species have a diameter that falls between 5 and 15 micrometers. Most spores have some type of ornamentation (Figure 2-16); it is questionable whether any are entirely smooth (Schoknecht and Small 1972). The various types of ornamentation that can be present on the spores of a particular species include **asperulate** (roughened or densely but very finely warted), **verrucose** (with scattered but relatively prominent blunt warts), **echinate** (with scattered but relatively prominent sharp-pointed spines), and **reticulate** (covered by a network of ridges). Both the size and the type of ornamentation are rather constant for a given species and are

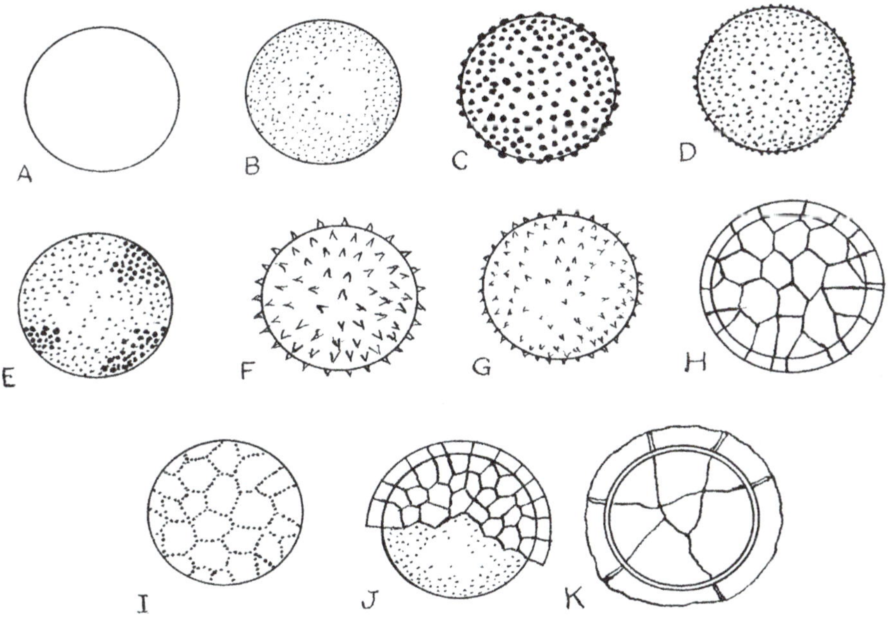

Figure 2-16. Types of spore ornamentation. (A) Smooth or nearly smooth. (B) Asperulate (roughened). (C) Verrucose (warted). (D) Verrucolose (delicately warted). (E) Verrucolose with clusters of more distinct warts. (F) Echinate (spiny). (G) Delicately spiny. (H) Banded reticulate. (I) Verrucose (or spiny) reticulate. (J) Partially reticulate. (K) Coarsely reticulate. Other types of ornamentation are possible (©Henry Stempen).

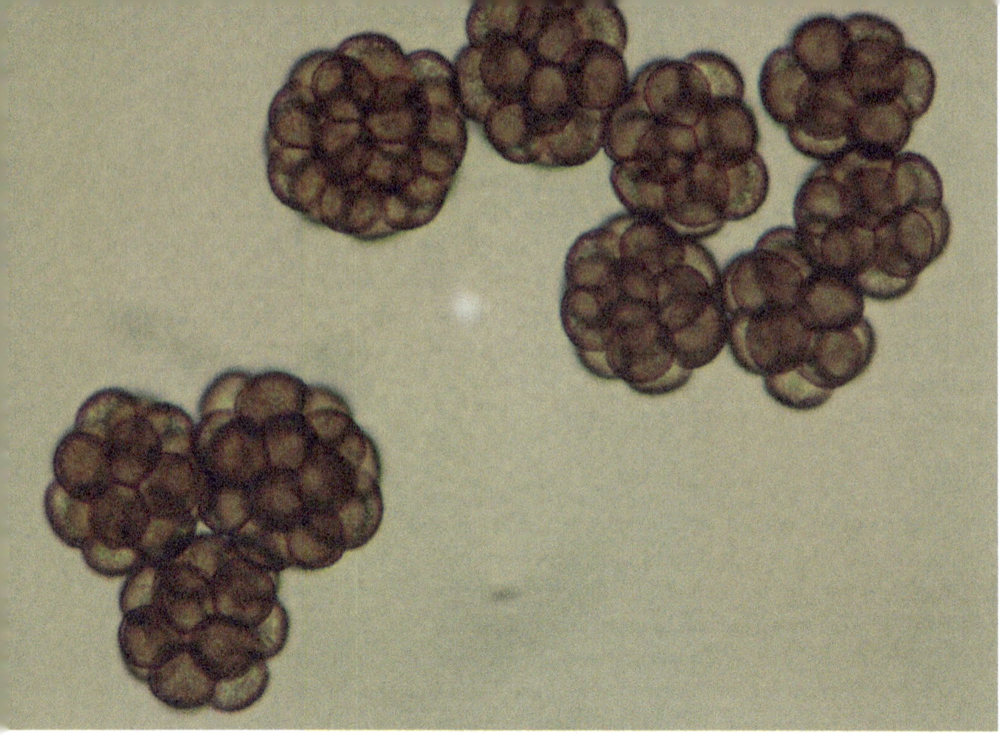

Figure 2-17. Clustered spores ("spore balls") of *Badhamia nitens* (630X) (©Alain Michaud).

often exceedingly important in identification. In fact, spore ornamentation is often the single most reliable morphological feature to separate closely related species that otherwise appear to be rather similar. However, this sometimes requires examining spores using electron microscopy. Spore size is not as constant a feature but still exceeding useful for identification. It should be noted that if spore size varies greatly for spores from the same fruiting body, it almost invariably indicates that the fruiting body did not develop normally.

Spore color in myxomycetes varies from hyaline to almost black; the only colors not found are blue and true green. Nevertheless, the spores of most species can be arbitrarily placed into one of two major color groups. Spores are either dark, as is the case for most members of the traditional orders Physarales and Stemonitales, or pale to brightly colored, which is the condition that exists for all other myxomycetes. The first group includes spores that appear brown, purplish brown, violet, or black, while the second group includes those that are white, pink, pale gray, light brown, rusty brown, yellow, orange, or red. These colors refer to the spores in mass as viewed in natural light. Spore color is much less evident when the spores are viewed by transmitted light under a microscope.

Spores are free in the vast majority of myxomycetes; that is, they do not adhere to one another. In a few species, however, such as *Dianema corticatum* and several members of the genus *Badhamia*, they occur in clusters known as **spore balls** (Figure 2-17).

CHAPTER 3

Collection, Culture, and Study of Myxomycetes

Collection of Myxomycetes in Nature

Collecting myxomycetes requires very little in the way of special equipment, although a sheath or pocket knife is useful for removing pieces of the substrate bearing fruiting bodies. Because of their fragile nature, specimens are best placed directly into small plastic boxes or other containers. The plastic tackle boxes with multiple compartments are ideal for this purpose. Particularly delicate specimens can be loosely wrapped in tissue paper to protect them from damage while being transported back from the field. If a plasmodium in the process of developing into fruiting bodies is encountered in the field, its location can be marked in some way so that the collector can return later in the same day to look for mature fruiting bodies. If the collector cannot return to the same spot, the specimen can be transferred to a box or can with a lid. A humid atmosphere is necessary for continued development, so if there is not enough moisture in the rotting substrate, a damp paper towel should be placed in the collecting container. It is important that the specimen not be flooded with water because this may hamper further development.

Upon returning from the field, the boxes or cans should be opened immediately. If the specimens consist of mature fruiting bodies, they should be placed aside and allowed to dry at room temperature. Most collectors then glue (using ordinary white glue) the pieces of the substrate bearing a particular specimen to the inside, either the bottom or the lid, of a small cardboard box for permanent storage (Figure 3-1). Some collectors prefer to glue the specimen (by its substrate) to a strip of stiff paper, such as a piece of index card, with the ends bent up. The strip, which is cut so that it fits snugly into the box, can be lifted out when the specimen is to be examined. This technique is especially appropriate when matchboxes or small pasteboard boxed are used as storage containers.

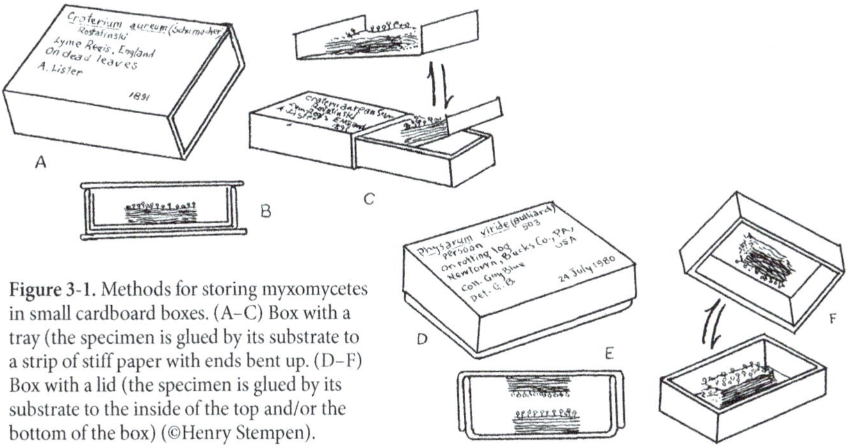

Figure 3-1. Methods for storing myxomycetes in small cardboard boxes. (A–C) Box with a tray (the specimen is glued by its substrate to a strip of stiff paper with ends bent up. (D–F) Box with a lid (the specimen is glued by its substrate to the inside of the top and/or the bottom of the box) (©Henry Stempen).

Each box containing a specimen should be labeled with such information as the species of myxomycete, the substrate upon which the fruiting occurred, the locality where the collection was made, the collection date, the collector's name and collection number, and the name of the person who identified the specimen if other than the collector.

The relatively small size of the storage containers used for myxomycetes permits a considerable number of specimens to be stored in a single drawer, small cabinet, or large plastic box (e.g., the plastic boxes used for storing shoes). Any of these will probably prove of sufficient size to accommodate most collections. If handled carefully, properly stored myxomycete fruiting bodies will remain suitable for study for many years. Moreover, myxomycete fruiting bodies are fairly resistant to most of the insects and other pests that often plague collections of preserved biological specimens. The chances of having damage caused by such organisms can be reduced by storing freshly collected specimens in the freezing compartment of a refrigerator for several days.

Relatively few biologists have studied the myxomycetes to any real extent, and distributional data are rather limited for some regions of the world. Consequently, someone who is so inclined could make a real contribution to science with some diligent collecting.

The intricate beauty of some myxomycete fruiting bodies may very well suggest a possible project for individuals interested in photography, although their small size makes such a project a real challenge. Nevertheless, excellent photographs indicating what can be accomplished have been featured in several magazine articles. Among these are articles by Barron (1991), Sharnoff (1991), and Johnson (2021). Moreover, myxomycetes have often been depicted in full color paintings or sketches, some of which are exceptionally beautiful. Examples include those found in Crowder (1926), Emoto (1977), and Mele (2020).

Moist Chamber Cultures

Throughout the temperate regions of the world, the fruiting bodies of myxomycetes tend to be relatively abundant only during summer and early autumn. During the remainder of the year, fruitings are encountered only occasionally, and these often are old specimens that have persisted in nature for several weeks or even months. The fruiting bodies of most myxomycetes are quite fragile and, as a result, tend to be ephemeral structures. Nonetheless, studies of myxomycetes do not have to be restricted to material collected in the field, since a relatively simple technique can be used to obtain both plasmodia and fruiting bodies in the home or laboratory. This technique, which involves placing pieces of tree bark or other types of organic debris in moist chamber cultures, is easy to carry out and requires no special equipment. Moreover, it can be used at any time of the year and in any part of the world (Stephenson 1985; Stephenson and Stempen 1994; Keller et al. 2021).

The use of moist chambers for culturing myxomycetes was first described by Gilbert and Martin (1933), who placed a few pieces of bark bearing an abundant growth of the green alga *Protococcus* in a moist chamber in their laboratory to permit the alga to develop. Much to their surprise, the fruiting bodies of a myxomycete also appeared on the bark in the moist chamber. This suggested further examination of similar bark cultures, and the two biologists soon discovered that the appearance of myxomycetes in such cultures was a common occurrence. Moreover, the species obtained in this way included some examples previously thought to be exceedingly rare. Since that time, the moist chamber culture technique has been widely used to supplement field collections of myxomycetes.

Collecting Bark

The first step in setting up a moist chamber culture involves collecting several postage stamp-sized or smaller pieces of the dead outer bark of a living tree. Various other types of organic debris, including dead leaves, twigs, and the dung of herbivorous animals, also may be used to prepare moist chamber cultures but are generally less productive than bark. The myxomycetes obtained from these other substrates will be discussed in the next chapter. Bark from some loose-barked trees may be collected by simply prying off pieces by hand, but for most trees it is necessary to remove pieces of bark with a knife or other collecting tool. A small screwdriver is both effective and safe to use.

When collecting pieces of bark, care should be taken not to disturb the living tissues of the tree. Inflicting damage has the potential effect of allowing disease-causing organisms to enter the tree. In addition, pieces of bark which include living tissue, when placed in moist chambers, are often quickly overgrown by filamentous fungi and usually have to be discarded.

Trees with smooth bark are generally regarded as less productive for myxomycetes than trees with rough bark, such as oak (*Quercus*), ash (*Fraxinus*), elm (*Ulmus*), maple (*Acer*), and hickory (*Carya*), which usually produce good results. In general, bark from coniferous trees tends to yield fewer species of myxomycetes than does bark from broadleaf trees, although some exceptions do exist. For example, bark from species of juniper (*Juniperus*) and wattle (*Acacia*) is often very productive. Various species of trees that are members of the same genus may exhibit a range of different bark types. For example, this is the case for the eucalypts (*Eucalyptus*) in Australia.

The pieces of bark taken from a given tree should be placed in a single collecting bag and the date, location, and type of tree recorded in pencil either on the bag itself or on a slip of paper placed in the bag with the bark. Plastic sandwich bags are suitable for short-term storage, but small paper bags should be used if the bark is to be stored for more than a few days.

Preparing Moist Chambers

The next step in setting up a moist chamber involves preparing the culture chamber itself (Figure 3-2). Petri dishes have been widely used as culture chambers, but any similar container or shallow bowl that can be covered with a top or lid will work just as well. Many modern workers use plastic Petri dishes, which are simply discarded when no longer needed. Plastic containers have another major advantage over glass containers. In a moist chamber, the fruiting bodies of myxomycetes sometimes develop on the side of the container and on the lower surface of the top or lid. When this happens, the only way to harvest the fruiting bodies is to cut away the portion of the container upon which they developed. This is rather difficult to do with a glass container, but relatively easy with plastic.

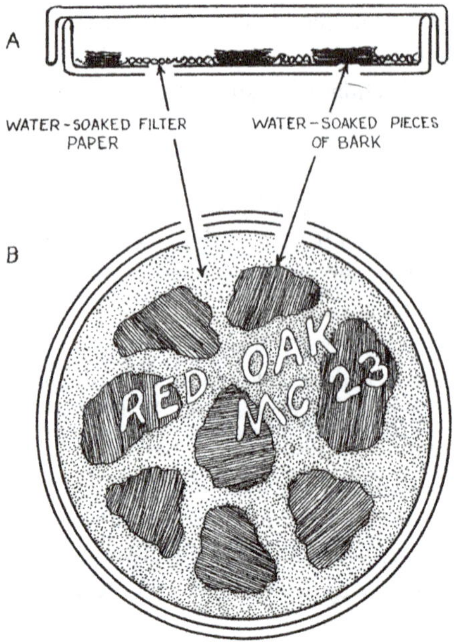

Figure 3-2. Petri dish set up as a moist chamber culture. (A) Side view. (B) Top view, showing the number assigned to the culture and the type of tree from which the bark was obtained recorded on the lid with a wax pencil or marking pen (©Henry Stempen).

Once a suitable container has been selected, it should be fitted with a filter paper disk or piece of absorbent paper towel trimmed to the appropriate size. For experiments designed to study the differences that exist in the species of myxomycetes obtained from bark samples from different tree species or trees growing in different habitats, care should be taken to prevent the potential contamination of one culture with spores from another by sterilizing all the equipment used in preparing moist chamber cultures, including the paper used to cover the bottom of the culture chambers. If the only objective of preparing cultures is simply to obtain a variety of different myxomycetes, sterilization is not necessary.

After the filter paper or paper towel is in place, it should be covered with samples of the organic debris (bark, leaves, twigs, or dung) to be cultured. The samples should not overlap each other to any great extent, and bark samples should be placed in the container with their cut surfaces facing downward. Enough distilled water (tap water can be used if distilled water is not available) is then added to completely cover the samples. The container is covered and set aside to allow the sample to soak up the moisture overnight.

Each container should be labeled, either with a piece of tape applied to the side so as not to obscure any portion of the bottom when the container is viewed from above or by writing directly on the lid with a wax pencil. If the latter method is used, it sometimes becomes necessary to rotate the lid of the dish when making observations of the samples to view portions of the bottom otherwise obscured by the writing. The next day, excess water in the container should be poured off and the cultures set aside and disturbed as little as possible. If Petri dishes are used, they may be stacked to save space. Cultures should be incubated at ordinary room temperatures under diffuse light.

Checking the Cultures

Cultures should be observed a day or two after they have been established and thereafter on a regular basis (that is, at least once every few days) for at least two weeks. After this time, cultures may be checked at less frequent intervals for as long as they are maintained. It is possible to maintain cultures for several months or more. In fact, some species of myxomycetes that appear in moist chamber cultures typically require weeks or even months to develop. This is almost always the case for species associated with dung. In contrast, many of the species that produce very small fruiting bodies, such as species of *Echinostelium* (including the exceedingly common *E. minutum*), often appear rather quickly, sometimes within 24 to 48 hours. Since the cultures slowly dry out, it may be necessary to add more water from time to time to keep cultures moist throughout the incubation period.

It is very important that cultures be checked as carefully as possible to detect all the fruiting bodies that may be present. Sometimes only one or a few fruiting bodies of a given species develop; in other cases hundreds of fruiting bodies are produced. Moreover, it is not at all unusual for a single culture to yield the fruiting bodies of several different species.

Cultures are most easily checked using a dissecting microscope, but a good hand lens (at least × 10) may be used if a microscope is not available. In some instances, fruiting bodies are clearly visible to the naked eye. Since removing the lid can interfere with the development of fruiting bodies not yet completely mature, a culture should be examined by viewing through the lid whenever possible.

When fruiting bodies are detected in a culture, they should be harvested as soon as circumstances permit. If this is not done, the fruiting bodies soon become moldy. Cultures should be left covered until the fruiting bodies are completely mature, although often it is helpful to raise the lid on one side to allow gradual drying of the fruiting bodies. When dry, the fruiting bodies can be harvested by removing both the fruiting bodies and portions of the substrate upon which they developed. Once the fruiting bodies have been removed from the culture, they should be allowed to air-dry. Then each individual collection should be placed in a small box for permanent storage, as described earlier in this chapter.

The question has been raised whether the species of myxomycetes appearing on various substrates placed in moist chamber cultures necessarily produce fruiting bodies on the same substrates in nature, since their appearance in such cultures may reflect nothing more than the presence of spores, microcysts, or sclerotia on these substrates (Brooks 1967; Keller and Brooks 1973; Blackwell and Gilbertson 1984). Although the data available on myxomycete distribution patterns are still too limited to permit a definitive answer to this question, there is mounting evidence that with diligent collecting many (and perhaps most) of the species obtained from moist chamber cultures prepared with material from a particular microhabitat occur as natural fruitings within the same microhabitat (Whitney 1982; Blackwell and Gilbertson 1984; Stephenson 1988, 1989). It does not always follow, however, that these species are necessarily restricted to the particular habitat, although this sometimes seems to be the case, as was reported by Arenas-Taborda et al. (2021) for a study carried out in Costa Rica.

Preparation of Material for Study

While it is possible to identify most genera and at least some of the more distinctive species of myxomycetes in the field, positive identification of most species requires examination of material under magnification in the home or laboratory. In many instances, the material must be mounted on a slide and examined with a compound microscope. Some of the techniques involved in preparing myxomycetes for microscopic study are described below.

To study the microscopic features of myxomycetes in some detail, it is necessary to acquire several small items of equipment and to prepare or obtain a number of chemical reagents. The single most essential item of equipment is a pair of fine pointed forceps, sometimes also called watchmaker's or jeweler's forceps. A good pair of forceps, which can be obtained from any biological supply house, is absolutely indispensable for handling individual fruiting bodies or transferring a portion of the spore mass and other associated structures from a given specimen to a microscope slide. In addition, it is exceedingly helpful to have at least two dissecting needles or probes for teasing portions of a fruiting body apart or dispersing spores and capillitium on a slide. Dissecting needles are supplied with most dissecting kits used in introductory biology courses and are also available from biological supply houses. An excellent probe can be made by simply inserting the blunt end of an ordinary sewing needle into a match stick or other small wooden handle.

Chemical reagents routinely used in studying the microscopic features of myxomycetes include distilled water, or, if such is not available, ordinary tap water, 95 percent ethanol, and a 3-percent aqueous solution of potassium hydroxide. The latter is prepared by adding 3 grams of potassium hydroxide to 97 milliliters of water. For convenience, these reagents are best placed in small dropper bottles.

The technique generally used to prepare material for microscopic examination involves transferring the specimen to be examined to a drop of 95-percent ethanol on a clean glass slide. After most of the ethanol has evaporated, a drop of potassium hydroxide is added. The addition of ethanol as an initial wetting agent serves to eliminate unwanted air bubbles from the material, while the potassium hydroxide causes the spores and other structures to expand to their full size. A cover slip is then placed on the slide and the latter is ready for examination under the microscope.

Slides prepared using the technique described above are for short-term study only. To produce semi-permanent slides, a mounting medium must be substituted for the potassium hydroxide. Two such media that are widely used for myxomycetes are Amann's lactophenol medium and Hantsch's fluid (Martin and Alexopoulos 1969). The former is prepared by combining 20 grams of phenol crystals, 20 grams of lactic acid, 40 milliliters of glycerine, and 20 milliliters of water. Because phenol is somewhat toxic, one should use extreme care when mixing or handling Amann's lactophenol medium. The addition of a small amount (no more than 0.1 gram per 100 milliliters of lactophenol is usually enough) of cotton blue to a bottle of Amann's lactophenol medium yields a stain that is often exceedingly useful for bringing out markings on hyaline or light-colored spores and capillitium, but it should not be used for dark-colored spores and capillitium, since it tends to obscure such markings. Hantsch's fluid consists of 30 milliliters of 90-percent ethanol, 20 milliliters of water, and 10 milliliters of glycerine. If necessary, mounts prepared with these two media can be heated gently to hasten the expansion of material to its full size. To make such slides semi-permanent, the edge of the cover slip should be sealed with clear fingernail polish or one of the slide sealing compounds available from biological supply houses. If prepared properly, such slides are virtually permanent. Each slide can then be labeled with a small gummed label or a glass-marking pen (Farr 1981).

It should be noted that Amann's lactophenol medium cannot be used to prepare mounts of myxomycetes with lime present in the fruiting body (that is, members of the Physarales). The reason for this is that the lactic acid in the medium reacts with the lime to produce bubbles of carbon dioxide, which disrupt or obscure the material on the slide. As such, Amann's lactophenol medium may be used as a quick test for the presence of lime, although a better test for lime is simply to add a drop of 5-percent lactic acid to the material. Unfortunately, there is no ideal mounting medium for preparing permanent slides of fruiting bodies with lime present. Nevertheless, it is possible to mount such material in a solution of 100 milliliters of a white corn syrup such as Karo, 300 milliliters of water, and 4 milliliters of 40-percent formalin (Stevens 1974). Spores mounted in this medium tend to shrink, but any structures consisting of lime, such as lime nodes of the capillitium, are well preserved.

When identifying species in such genera as *Collaria*, *Badhamia*, *Comatricha*, *Paradiacheopsis*, *Physarum*, *Stemonitopsis*, and *Stemonitis*, the arrangement and basic structure of the capillitium are exceedingly important diagnostic features which are best observed in fruiting bodies from which most of the spores have been dispersed. If all the fruiting bodies in a given collection are examined under a dissecting microscope, it is usually possible to find several from which enough spores have been lost to reveal the nature of the particular structure, whether pseudocolumella, columella, or capillitium. If no such examples are available, it is often possible to disrupt the spore mass of an intact fruiting body by first removing it with the fine-pointed forceps and then gently blowing to displace the spores. For members of the traditional order Stemonitales, repeatedly dipping the fruiting body into a large drop of ethanol on a slide is also an effective method of displacing spores.

As noted in Chapter 2, a microscope equipped with an ocular micrometer and an oil immersion objective is a prerequisite for the accurate determination of some features. One of the most important of these features is spore size, which can be determined only with the aid of a microscope having an ocular micrometer mounted in the lens tube of the eyepiece. An ocular micrometer is a small, round piece of glass with a measuring scale permanently etched on it.

Before the measuring scale on the ocular micrometer can be used, the units on it have to be calibrated or compared to a known standard to determine what each unit on the scale represents. This requires the use of a stage micrometer and should be done for each objective lens (and thus each magnification) of the microscope. The technique used to calibrate a microscope is described in many introductory laboratory manuals as well as in most textbooks on microtechnique and microscopy (Farr 1981). A fairly complete description of the technique is also available in Largent et al. (1977) and Smith et al. (1981).

CHAPTER 4

Distribution of Myxomycetes Around the World

Myxomycetes are common inhabitants of decaying plant material throughout the world. They are particularly abundant in forested regions, where decaying logs, stumps, and dead leaves furnish a plentiful supply of potential substrates, but it is probably safe to say that at least some species occur in any terrestrial ecosystem in which plants are present. Although myxomycetes are regarded as terrestrial organisms, at least a few species sometimes occur in aquatic habitats. Shearer and Crane (1986) recorded 13 species on submerged decayed plant substrates and balsa wood baits from two swamps in southern Illinois in the United States, Lindley et al. (2007) isolated 14 species from samples of plant debris collected from ponds and lakes in Arkansas and Oklahoma, and Gottsberger and Nannenga-Bremekamp (1971) described a new species of presumably "aquatic" myxomycete (*Didymium aquatile*) from Brazil. The plasmodial stage of *D. aquatile* apparently lives submerged in water until just before fruiting. They reported that plasmodia of this species were observed creeping along on the sandy bottom of a stream, sometimes at a depth of 10 to 20 centimeters. Although *D. aquatile* would seem to represent an extraordinary example, plasmodia of other species of myxomycetes have been grown under water in culture (Ward 1886; Parker 1946), and one of these (*Didymium difforme*) was reported as being capable of completing all stages of its life cycle under water (Ward 1886). Moreover, it is not uncommon to observe myxomycete plasmodia submerged on the inner surfaces of glass vases holding cut flowers (M. Blackwell, personal communication).

Geographical Distribution

Most species of myxomycetes appear to be cosmopolitan, but some are known to have a more restricted distribution. At least a few species seem to be confined to the tropics or subtropics, and a number of others have been collected only in temperate regions. Two well known examples of temperate myxomycetes are the relatively common *Hemitrichia clavata* and *Leocarpus fragilis*, whereas tropical myxomycetes include such species as *Ceratiomyxa sphaerosperma*, *Craterium paraguayense*, and *Alwisia bombarda*.

As is true for most groups of organisms, what we know about the distribution of myxomycetes is based upon species lists that have been compiled by collectors for particular regions of the world. The axiom that the known distribution of an organism tends to be highly correlated with the distribution of the people who collect or study it certainly applies to the myxomycetes; some regions are much better known than others simply because of the presence of individuals interested in myxomycetes. Areas of North America for which published lists are available include Big Bend National Park (Ndiritu et al. 2009a), Everglades National Park (Keller 1973), Great Smoky Mountains National Park (Welden 1951; Stephenson et al. 2001), the Long Island area of New York (Hagelstein 1936), the Sonoran Desert of Arizona (Blackwell and Gilbertson 1980), the Canadian province of Alberta (Richardson and Currah 1990), and such states as Arkansas (Eliasson et al. 1988), California (Critchfield and Demaree 1991), Colorado (Mitchel et al. 1980), Hawaii (Eiasson

1991), Louisiana (Lowy 1953), Mississippi (Russell and Huneycott 1972), Ohio (Fullmer 1921), Texas (Alexopoulos 1965; Alexopoulos and Henney 1971), West Virginia (Stephenson 1982), and Wyoming (Morris 1954) in the United States. Other areas of the world for which considerable data exist include Argentina (Wrigley de Basanta et al. 2010; Lado et al. 2011), Australia (Mitchell 1995; Knight and Brims 2010; Wellman 2016, 2017; Stephenson 2021), Brazil (Farr 1968), the British Isles (Ing 1980, 1982, 1999), Chile (Lazo 1966; Lado et al. 2007, 2013), China (Li and Li 1989), Columbia (Uribe-Meléndez 1995), Costa Rica (Alexopoulos and Sáenz 1975; Rojas et al. 2018), Ecuador (Farr et al. 1979; Schnittler et al. 2002), Ethiopia (Dagamac et al. 2017), Finland (Härkönen 1979, 1981a, 1981b, 1989), Greece (Alexopoulos 1959), Greenland (Gøtzsche 1989), Iceland (Gøtzsche 1984, 1990), India (Lakhanpal and Mukerji 1981; Ranade et al. 2012), Israel (Binyamini 1997), Kazakhstan (Schnittler et al. 2001a, Zenlyanskaya et al. 2020), Laos (Ko Ko et al. 2012), Mexico (Braun and Keller 1976; Keller and Braun 1977; Lado et al. 2003; Estrada-Torres et al. 2009), Myanmar (Ko Ko et al. 2013), Nepal (Yamamoto and Hagiwara 1990), New Zealand (Mitchell 1992; Stephenson 2003), Norway (Johannesen 1984), Peru (Lado et al. 2016), the Philippines (Macabago et al. 2012), Puerto Rico (Novozhilov et al. 2000), Spain (Illana et al. 1990), Sweden (Santesson 1964), Tanzania (Härkönen and Saarimäki 1991; Ukkola et al. 1996), Thailand (Ko Ko et al. 2010), and Vietnam (Novozhilov et al. 2017). A few multi-country checklists have been compiled, including Lado (1994) for the Mediterranean, Lado and Wrigley de Basanta (2008) for the Neotropics, and Ndiritu et al. (2009b) for the countries of Africa. Isolated islands that have been surveyed for myxomycetes include Ascension Island (Stephenson 2009), the Canary Islands (Beltran et al. 2004), Christmas Island (Stephenson and Stephenson 2019), Cocos Island (Rojas and Stephenson 2008), La Reunion Island (Adamonyte et al. 2011), Norfolk Island (Stephenson and Stephenson 2020), Macquarie Island (Stephenson et al. 2007), and the Seychelles Islands (Kryvomaz et al. 2020).

A comparison of the lists compiled for the different areas reveals that they invariably include such species as *Arcyria cinerea*, *A. denudata*, *Ceratiomyxa fruticulosa*, *Fuligo septica*, *Lycogala epidendrum*, *Stemonitis fusca*, and *Stemonitopsis typhina*. As such, it would seem apparent that these are among the most consistently abundant and widely distributed myxomycetes in the world. Nevertheless, although many of the more common species of myxomycetes are among the most ubiquitous of organisms, both the assemblage of species present and their relative abundances can be markedly different from one place to another. In fact, it is possible for species that are exceedingly abundant in one region to be rare or uncommon in another region. For example, one is much more likely to encounter *Amaurochaete ferruginea* in the forests of the Rocky Mountains than elsewhere in North America, and certain members of the traditional order Physarales, such as *Lepidoderma trevelyani* and *Physarum bitectum*, appear to be generally more abundant in western North America than in eastern North America.

Some of these differences are probably related to the fact that the forests over most of eastern North America consist largely of broadleaf trees (Figure 4-1), while the forests throughout much of western North America are primarily coniferous (Figure 4-2). Although many myxomycetes do not appear to show any preference for substrates derived from either coniferous or broadleaf trees, this is not the case for at least some species. For example, *Cribraria cancellata* and most other species of *Cribraria* are more common on decaying wood from coniferous trees than wood from broadleaf trees. In the Appalachian Mountains

Figure 4-1. Broadleaf forest in the eastern United States (©Steve Stephenson).

Figure 4-2. Coniferous forest in central Alaska (©Steve Stephenson).

of eastern North America, where coniferous forests generally occur at higher elevations and broadleaf forests at lower elevations, the differences that exist for the assemblages of myxomycetes associated with the two types of forests are sometimes quite apparent. For example, such species as *Cribraria rufa*, *Lepidoderma tigrinum*, *Licea minima*, *Stemonitopsis aequalis*, and *Trichia erecta* are relatively common in the high-elevation coniferous forests but are rarely encountered in forests at lower elevations (Stephenson 1983, 2004).

Since the forests of eastern North America and Europe are fairly similar, one might expect the assemblages of myxomycetes associated with these forests to be rather comparable. For the vast majority of the more common and widely distributed species, this does indeed seem to be case. At least some species, however, appear to be more common in one region than in the other, and a few species are known only from one of the two regions. For example, *Diachea thomasii* and *Tubifera microsperma* are apparently largely restricted to North America, while *Brefeldia maxima*, *Dianema corticatum*, and *Trichia verrucosa* are much less common in North America than in Europe (Martin and Alexopoulos 1969).

Stephenson et al. (2000, unpublished data) compiled all readily available digitized records of myxomycetes for the eastern and western sections of the contiguous United States into a single database for each of the two sections. The database for the eastern United States consisted of more than 58,000 records representing a total of 460 species, while the database for the western United States consisted of more than 24,000 records representing 419 species. When the two databases were combined, one or more records existed for 531 different species, and 399 of these were shared in common between the two sections of the contiguous United States. Some species, including such common examples as *Metatrichia vesparia* and *Trichia varia* were noticeably more abundant in one section or the other.

Myxomycetes are not equally abundant in all types of terrestrial ecosystems. For example, they appear to be much less abundant in tropical forests (Figure 4-3) than in temperate forests (Alexopoulos 1970; Farr 1976). A number of reasons have been advanced to account for this difference in abundance. Among these are the excessive acidity of many of the substrates available to myxomycetes in tropical forests, the lack of air movement for myxomycete spore entry and dispersal, daily torrential rainfall that washes most developmental stages of myxomycetes off the substrates upon which they typically occur, high humidity which promotes the invasion of fruiting bodies by filamentous fungi, large numbers of small animal predators (e.g., insects, arachnids, and slugs) that feed on myxomycete spores, and a light intensity that is too low to allow normal development of fruiting bodies in those species of myxomycetes for which light is a controlling factor (Alexopoulos 1970; Farr 1976; Maimoni-Rodella and Gottsberger 1980). It has also been suggested (Blackwell 1984; Wheeler 1984a) that the species of myxomycetes occurring in tropical forests may exist mostly as plasmodia and only occasionally form fruiting bodies. If such is the case, the apparent sparseness of tropical myxomycetes is perhaps more illusory than real.

Figure 4-3. Tropical forest in northern Thailand (©Steve Stephenson).

Snowbank Species

One group of myxomycetes is restricted to alpine areas such as the Rocky Mountains of western North America, the Alps of Europe, and the Snowy Mountains of Australia (Stephenson and Shadwick 2009), where its members are found fruiting along the margins of melting snowbanks in the late spring and early summer (Figure 4-4). The species that occupy this rather special and very limited habitat are usually referred to as **snowbank** or **snowmelt myxomycetes**. They constitute a distinct ecological group, since they usually produce fruiting bodies only during the relatively brief period of time when the special microenvironmental conditions associated with melting snowbanks and required for their growth and fruiting exist. During the remainder of the summer, the species of myxomycetes found in these alpine areas are very much the same as those collected at lower elevations in the same regions.

Figure 4-4. Snowbank habitat in the Snowy Mountains of Australia (©Steve Stephenson).

Among the most common and widely distributed myxomycetes associated with snowbank myxomycetes are *Arcyria versicolor*, *Diderma niveum*, *Lamproderma sauteri*, *Prototrichia metallica*, and *Trichia alpina*. *Arcyria versicolor* and *P. metallica* often form extensive fruitings on decaying logs and other types of woody debris shortly after these are exposed by a melting snowbank. *Diderma niveum* and *L. sauteri* are more likely to be encountered on litter or the leaves and stems of low-growing shrubs and other living plants. Other characteristic species of alpine areas include *Comatricha suksdorfii*, *Hemitrichia montana*, *Lepidoderma chailletii*, *Meriderma carestiae*, and *Physarum alpinum*. In general, the fruiting bodies of snowbank species tend to be more robust than the fruiting bodies of other myxomycetes, and alpine fruitings of species also occurring in lowland habitats are characterized by spores that are somewhat larger than typical spores. The significance of this phenomenon is unknown. Interestingly, the majority of the species in some genera tend to be predominantly alpine in distribution. This is true for *Dianema* (Kowalski 1967), *Lamproderma* (Kowalski 1970), and *Lepidoderma* (Kowalski 1971).

Snowbank myxomycetes occur in alpine areas at higher elevations throughout the temperate zone (Tamayama 2000), and some members of the group have even been reported from alpine zones of tropical mountains (Farr 1976). Snowbank myxomycetes appear to be poorly represented in high-latitude regions of the world, and several collections of *Diderma niveum* from southeastern Alaska (58°N latitude) apparently represent the northernmost records of this group in North America (Saccardo et al. 1904). In Europe, *Diderma niveum*, *Lamproderma sauteri*, and *Lepidoderma carestianum* are known to occur as far north as 64°N latitude (Gøtzsche 1984).

High-Latitude Species

Although a few species, including *Lycogala epidendrum*, *Oligonema schweinitzii*, and *Trichia subfusca*, have been collected at Utqiagvik (71°N latitude) in extreme northern Alaska, myxomycetes do not appear to be particularly common in such high-latitude areas of North America. Stephenson and Laursen (1993) recorded only 17 species from arctic and subarctic alpine tundra of central and northern Alaska. *Mucilago crustacea* is perhaps the most common and undoubtedly the most conspicuous myxomycete associated with arctic tundra. Active plasmodia and aethalia of this species are not uncommon during late July and early August, and it is not unusual to encounter several fruitings in an area of no more than 100 square meters. Other species recorded from either arctic or subarctic alpine tundra in Alaska include *Fuligo intermedia*, *Perichaena chrysosperma*, and *Trichia favoginea*. The numbers of species recorded for several high-latitude areas of Europe are considerably higher. For example, 32 species are known from Iceland (at 64 to 66°N latitude) and 54 species have been reported from various localities (60 to 77°N latitude) in Greenland (Gøtzsche 1989, 1990).

There have been only a few reports of myxomycetes from high-latitude areas of the Southern Hemisphere. In what apparently represents the first record of a myxomycete from the south polar region, Horak (1966) described a new species, *Diderma antarcticolum*, from material collected on the Antarctic Peninsula (64°S latitude). Later, Ing and Smith (1980, 1983), Arambarri and Spinedi (1989), and Stephenson et al. (1992) reported a number of additional species, including *Diderma niveum*, *Didymium dubium*, and *Lamproderma ovoideum*, from both the Antarctic Peninsula and several of the subantarctic islands. A checklist of all records of myxomycetes known from Antarctica was provided by Thompson et al. (2019).

Desert Species

Since myxomycetes are almost invariably associated with relatively cool and moist habitats, one might not expect these organisms to occur in deserts (Figure 4-5). However, the number of species reported from desert regions such as those of the southwestern United States, Mexico, and South America is surprisingly high. Blackwell and Gilbertson (1980) reported 33 species from the Sonoran Desert of southern Arizona, bringing the total number of known species for this region of North America to 46. Since then, other more extensive studies (e.g., Lado *et al.* 1999, 2016; Wrigley de Basanta *et al.* 2008; Estrada-Torres *et al.* 2009) carried out in other deserts of the Americas have reported even higher totals. These studies have also demonstrated that there is a distinct assemblage of myxomycetes associated with native cacti and other succulent plants (e.g., certain members of the families Agavaceae, Bromeliaceae, and Euphorbiaceae) in desert habitats. Lado *et al.* (2007) listed ten species of myxomycetes that were described as new to science from specimens associated with succulent plants. Additional species of what are referred to as **succulenticolous** myxomycetes have been described since then (e.g., Estrada-Torres *et al.* 2009; Lado *et al.* 2013). Some of these succulenticolous myxomycetes are not yet known from any other substrate. The most productive substrates for myxomycetes in deserts include the pith skeletons of decaying cacti, the dead parts of living plants in contact with the ground, animal dung, and the bark of living trees (Evenson 1961; Blackwell and Gilbertson 1980, 1984). Field collections are usually limited to no more than a few days or weeks immediately following a period of significant precipitation, but myxomycetes are exceedingly common in moist chamber cultures prepared with desert plant material. Some of the more common species in deserts are *Badhamia melanospora*, *Comatricha laxa*, *Didymium difforme*, *Didymium vaccinum*, *Perichaena depressa*, and *Physarum compressum*.

Figure 4-5. Desert in the southwestern United States (©Steve Stephenson).

Species Associated with Particular Microhabitats

The best-known myxomycetes are those characteristically associated with decaying wood or bark (Figure 4-6). The species found on such substrates often occur in great profusion, typically producing fruiting bodies of sufficient size to be easily detected in the field (Martin and Alexopoulos 1969). Not all decaying wood (also referred to as coarse woody debris) is the same, and the source of the wood (broadleaf trees versus coniferous trees), stage of decomposition, surrounding environmental conditions, and properties of the wood itself (e.g., pH) all have an influence on the assemblages of myxomycetes present (Fukasawa et al. 2015).

Figure 4-6. Decaying wood (coarse woody debris) as a microhabitat for myxomycetes (©Steve Stephenson).

Most of the myxomycetes occurring in temperate forests would seem to be **lignicolous**, since often more than 90 percent of field collections are made from the microhabitats represented by decaying wood and bark (Stephenson 1988). Nonetheless, a number of other microhabitats exist. Among these are the bark surfaces of living trees, leaf litter on the forest floor, aerial litter, and the dung of herbivorous animals. Until recently, the species of myxomycetes associated with these other microhabitats have received relatively little study. One reason for this is that many of the species involved tend to be rather inconspicuous or sporadic in their occurrence and thus difficult to detect in the field. In these cases, the moist chamber culture technique (described in Chapter 3) provides a convenient and often very productive method of supplementing field collections.

Figure 4-7. Bark surface of a living tree as a microhabitat for myxomycetes (©Steve Stephenson).

The Bark Surface of Living Trees as a Microhabitat

Except for the efforts of a few individuals such as the Rev. William Cran in Scotland (Lister 1938), myxomycetes occurring on the bark surface of living trees received little attention until the introduction of the moist chamber technique. Since that time, these **corticolous** myxomycetes have been the subject of numerous studies. Well over 100 species of myxomycetes now have been reported from the bark microhabitat as field and/or moist chamber collections (Mitchell 1980; Härkönen and Ukkola 2000; Jordan et al. 2006). Many of these are also known to occur in other microhabitats, but at least some species seem restricted to the bark of living trees. Prominent examples include various species of *Echinostelium*, *Licea*, and *Macbrideola* (Alexopoulos 1964; Brooks 1967; Mitchell 1980). Other myxomycetes commonly appearing on pieces of bark placed in moist chamber cultures include *Calomyxa metallica*, *Comatricha nigra*, *Enerthenema papillatum*, *Paradiacheopsis fimbriata*, and *Perichaena chrysosperma*.

Considerable variation exists for the bark microhabitats potentially available for corticolous myxomycetes. Different tree species exhibit a wide range of variation in bark surface physical characteristics (usually referred to as bark texture), with some trees exhibiting relatively smooth bark and others having furrowed and rather rough bark (Figure 4-7). Other characteristics known to vary from one type of tree to another are pH, water-holding capacity, and nutrient concentrations (Stephenson 1989; Härkönen and Ukkola 2000). For example, the bark of coniferous trees tends to be relatively more acidic than the bark of broadleaf trees. Studies have shown that the presence of epiphytic lichens and bryophytes on bark surfaces tends to increase the number of species of myxomycetes present, presumably because they trap myxomycete spores (which are largely wind-disseminated) from the air. Since the relative abundance of epiphytes can vary greatly among different types of trees, this factor would seem to have considerable potential for affecting the distribution patterns of corticolous myxomycetes. If it is hypothesized that myxomycete spores land with more or less equal frequency on all trees present at a particular site, then the differences in corticolous myxomycetes present are presumably the result of the variation that exists for tree bark.

Figure 4-8. Forest floor litter as a microhabitat for myxomycetes (©Steve Stephenson).

Litter as a Microhabitat

The fact that some species of myxomycetes seem to be associated with forest floor litter is well known (Gray and Alexopoulos 1968). Fruitings of such species as *Diderma effusum*, *D. testaceum*, *Didymium melanospermum*, *Lamproderma scintillans*, *Physarum bivalve*, and *P. cinereum* are collected almost exclusively from forest floor litter and thus are presumptively litter-inhabiting (Figure 4-8). Samples of leaf litter placed in moist chamber cultures often yield myxomycetes, but relatively little use has been made of this technique in studying the myxomycetes associated with the litter microhabitat. Nevertheless, on the basis of the few studies that have been carried out, it seems apparent that some species of myxomycetes display an affinity for leaf litter of either coniferous or broadleaf trees, while others give no indication of a preference for a particular type of litter. For example, Stephenson (1989) reported that *Stemonitis herbatica* apparently exhibits a preference for the litter of broadleaf trees, while *Cribraria microcarpa* seems to prefer coniferous litter. *Arcyria cinerea*, on the other hand, tends to appear in some abundance on all types of litter. In a study carried out in Finland, Härkönen (1981b) prepared a series of moist chamber cultures with fallen leaves collected from seven species of trees in several different localities. She found that certain types of litter invariably yielded particular myxomycetes, even though collected at different times and at different localities. For example, species of *Arcyria* developed only on the litter of coniferous trees, while species of *Didymium* were restricted to the litter of broadleaf trees. Other studies of the myxomycetes associated with litter include Stephenson et al. (1999), Tran et al. (2008), Rollins and Stephenson (2012), Goad and Stephenson (2013), Pelloquin and Stephenson (2016), and Nguyen et al. (2019).

However, the patterns of distribution of myxomycetes among different types of litter may involve certain less apparent factors. Rojas et al. (2020) collected samples of dead leaves from the forest floor in a temperate forest in Arkansas and a tropical forest in Costa Rica. These samples were sterilized and then placed in small-mesh nylon bags. Half of the samples from Arkansas and Costa Rica were placed out in the field in the same study area where they were collected and the other half were placed out in the other type of forest. After being left out in the field for four months, the samples were recollected and processed using moist chamber cultures. Samples left out in Costa Rica yielded appreciably fewer records and species than did the samples left out in Arkansas. These data suggest that environmental factors may play a major role in determining just which species of myxomycetes colonize litter.

Until recently, the concept of litter was limited to the dead leaves and other types of plant debris found on the ground surface (or forest floor, since the litter layer is best developed in forests). However, it is now recognized that there is a type of litter in forests that occurs above the ground. This "aerial litter" microhabitat consists of dead but still attached plant parts that have never been in contact with the ground. There is relatively little aerial litter in temperate forests, but it is often abundant in tropical forests (Novozhilov et al. 2021; Stephenson and Stephenson 2022). In some instances, it had been reported that moist chamber cultures prepared with samples of aerial litter are more productive than those prepared with samples of ground litter collected at the same locality (Black et al. 2004).

Interestingly, aerial litter is not the only type of aerial microhabitat. Lianas (vines) are not particularly common in temperate or boreal forests but can be abundant in tropical forests (Coelho and Stephenson 2012). Myxomycetes are surprisingly abundant on tropical lianas, both as specimens collected in the field and those appearing in moist chamber cultures (Wrigley de Basanta et al. 2008). Among the species associated with lianas are *Didymium squamulosum* and *Physarum pusillum*. Two other aerial microhabitats in the tropics are the inflorescences of giant herbaceous plants such as those produced by members of the genus *Heliconia* (Schnittler and Stephenson 2002) and the epiphyllous liverworts and lichens that form a covering over the surface of living leaves of woody plants (Schnittler 2001).

Aerial litter represents the source of an entirely different and rather unusual microhabitat. When they become detached, dead leaves and other plant parts can fall into the water of ponds and streams, where decomposition ultimately will take place. However, when some of this submerged plant material is collected, allowed to dry out completely, and then used to prepare moist chamber cultures, it is not uncommon for the cultures to yield the fruiting bodies of myxomycetes (Lindsey et al. 2007; Winsett and Stephenson 2013). This "submerged litter" microhabitat exists throughout all regions of the world but has only rarely been studied.

Dung as a Microhabitat

It has long been known that some myxomycetes fruit on the weathered dung of herbivorous mammals (Lister 1918), but relatively few studies (Merrill 1969; Eliasson and Lundqvist 1979; Cox 1981; Onduso et al. 2019; Stephenson and Novozhilov 2021) of **coprophilous** myxomycetes have been carried out (Figure 4-9). Nevertheless, more than 120 species in at least 23 genera have been recorded from dung, and a few of these seem to occur predominantly (or even exclusively) on this substrate (Eliasson and Lundqvist 1979; Eliasson 2013; Calaça et al. 2020).

Figure 4-9. Dung as a microhabitat for myxomycetes (©Steve Stephenson).

Dung is a highly complex substrate (Hudson 1986), and the exact role of myxomycetes in the communities of microorganisms associated with its decomposition is not yet known

(Eliasson and Lundqvist 1979; Madelin 1984). In any case, dung has several characteristics that seemingly make it unusually favorable for myxomycetes, including a high moisture content, large microbial populations, and nutrient richness (Hudson 1986). For example, the nitrogen content of dung (sometimes as high as 4 percent) is many times higher than for the woody substrates on which lignicolous myxomycetes occur. Dung also has a much higher pH than do most other substrates upon which myxomycetes are typically found (Emoto 1938; Stephenson 1988). Presumably, this is reflected in the pH optima for those species that are truly coprophilous.

Although distributional data for the group are exceedingly limited, it would appear that coprophilous myxomycetes are not particularly abundant in either temperate forests or tropical forests. Presumably, this is because the rapid degradation of dung by arthropods does not allow the myxomycetes time enough to become established (Novozhilov et al. 2021). For example, in a study carried out in a temperate deciduous forest in southwestern Virginia in the United States, Stephenson (1989) recorded only five species on dung, and none of these was considered truly coprophilous. Moreover, only one of the species (*Arcyria cinerea*) occurring on dung was represented by more than a single collection.

In contrast to their apparent low level of abundance in temperate and tropical forests, coprophilous myxomycetes would seem much more common in grasslands and deserts. For example, in a study of the coprophilous fungi (but not specifically myxomycetes) associated with different kinds of dung in a grassland in northeastern Colorado in the United States, Angel and Wicklow (1975) recorded three species of myxomycetes (*Didymium squamulosum*, *Kelleromyxa fimicola*, and *Licea tenera*), all of which are considered coprophilous or are commonly reported from dung. A series of cultures prepared with American bison (buffalo) dung collected in a grassland in northwestern Montana yielded numerous collections of such species as *Didymium anellus*, *D. difforme*, *Kelleromyxa fimicola*, *Perichaena depressa*, and *P. liceoides*) (Stephenson and Novozhilov 2021). In one study carried out in the southwestern United States, 20 different species, including several considered coprophilous, were reported from the dung of various desert animals (Blackwell and Gilbertson 1984). As a general rule, myxomycetes rarely occur, in the field or in moist chamber cultures, on the dung from the smallest mammals such as mice and squirrels. Dung from larger mammals such as cows and horses is the most productive.

Soil as a Microhabitat

Myxomycetes have been recorded on, or isolated from, soil by a number of workers, including Thom and Raper (1930), Warcup (1950), and Indira (1968), but only recently has it been demonstrated that these organisms actually occur in some abundance in soil. Feest and Madelin (1985, 1988a, 1988b) used a relatively simple enumeration technique to obtain data on the number of myxomycete plasmodium-forming units present in soil samples collected in England and elsewhere. Their data indicate that myxomycetes are rather common organisms in most soils, with the highest numbers generally being recorded from grassland and agricultural soils (Fiore-Donno et al. 2017). In temperate regions where myxomycetes have been studied most extensively, myxomycetes and dictyostelids are often the dominant protists (Urich et al. 2008), with the number of myxomycetes alone sometimes exceeding 20,000 cells per cm^3 (Feest and Campell 1986; Feest 1987). Based on this level of abundance, it has been suggested that myxomycetes are likely to spend the majority of their existence as amoebae in the soil. The ability of amoebae to alternate between two active forms and one inactive form (amoebae with or without flagella and resistant microcysts, respectively), depending on environmental conditions, and the ability to generate large clonal populations certainly represent major factors in their apparent dominance in soils. Little is known about the taxonomic composition of the assemblage of myxomycetes associated with the soil microhabitat (Stephenson et al. 2011; Stephenson and Feest 2012). However, data obtained with use of environmental polymerase chain reaction (PCR) suggest that genetic diversity is high, with species of *Didymium* and *Lamproderma* the most widespread and abundant (Fiore-Donno et al. 2016; Gao et al. 2019).

Canopy soil, as first used by Stephenson and Landolt (1998), refers to the layer of dead organic matter associated with the epiphytes that occur on the branches and (to a lesser extent) the trunks of trees in tropical rain forests. This dead organic matter is derived from decaying epiphytes, partially decomposed bark insect frass, and intercepted litter. It was first demonstrated to support dictyostelid slime molds, but additional studies revealed that canopy soil also supports myxomycetes (Stephenson and Landolt 2015), albeit at apparently low levels of diversity and abundance. The canopy soil microhabitat is found in temperate rain forests but data on the myxomycetes present are exceedingly limited.

CHAPTER 5

Ecology of Myxomycetes

Although brief notes relating to ecological aspects of myxomycetes have traditionally been included in taxonomic treatments of the group, predominantly ecological studies of these organisms are few. As a result, our knowledge of the ecology of myxomycetes is still rather limited, even though considerable advances have been made with respect to our knowledge of their geographical distribution, physiology, development, ultrastructure, and genetics.

Most of what is known about the ecology of myxomycetes is based upon observations of the reproductive, spore-producing stage in their life cycle. The reason for this is simply that myxomycetes are very seldom evident in nature except at this one point in their life cycle. The microscopic amoeboflagellate cells that represent the unicellular trophic stage in the life cycle are (at least for some species) relatively easy to study in the laboratory but impossible to observe under field conditions. The multinucleate trophic stage, the plasmodium, although often macroscopic, is also frequently cryptic and typically does not emerge to the surface of a substrate until ready to sporulate. Because myxomycetes tend to be directly observable only during the reproductive stage of their life cycle, one can only assume that the abundance of fruiting bodies at a particular locality or in a given microhabitat is correlated with the abundance of one or both trophic stages. Obviously, it is certainly possible that this is not always the case.

Trophic Stages

Both trophic stages in the myxomycete life cycle characteristically occur in microhabitats where bacteria are plentiful and there seems little doubt that bacteria constitute an important (if not the major) source of food for these organisms (Madelin 1984). Little appears to be known about the range of bacteria on which the amoeboflagellate cells of myxomycetes feed, although several types of bacteria, such as *Escherichia coli* and *Enterobacter aerogenes*, have been used as food sources in laboratory studies. Some types, however, appear to be more favorable than others (Gray and Alexopoulos 1968). In addition to bacteria, amoeboflagellate cells feed on the spores of various fungi (Gilbert 1928), and the very fact that these cells can be grown on chemically defined nutrient solutions in laboratory cultures suggests that at least some direct absorption of dissolved organic nutrients occurs in nature.

Plasmodia have been shown to be capable of feeding on a wide variety of bacteria (Cohen 1941), but not all types of bacteria are equally suitable for maintaining vigorous plasmodial growth and a few types actually inhibit growth. Blackwell (1974) found that remnants of the bacterial cells present in food vacuoles of some species of myxomycetes were extruded into the stalk at the time of sporulation. Using transmission electron microscopy, she was able to detect bacterial spores and other organic debris in the stalks of seven species of myxomycetes. The material examined included both specimens obtained from laboratory cultures as well as specimens collected in the field. The food material contained in the stalks, however, could seldom be identified.

No additional data are available on the bacterial food sources actually exploited by plasmodia in nature (Blackwell and Gilbertson 1984), although various food items are consumed by plasmodia of at least some species of myxomycetes. Among these are yeasts, algal cells, and the spores, vegetative hyphae, and (less commonly) the fruiting bodies of fungi (Lister 1888; Howard and Currie 1932a, 1932b). In fact, a few species of myxomycetes, including *Physarum polycephalum* and *Badhamia utricularis*, can probably be regarded as primarily fungivorous. Interestingly, the feeding of one plasmodium on another of a different species is also known to occur (McManus 1962), and it is possible that such interactive relationships may be relatively important in some situations that might develop in nature, such as the co-occurrence of several species of myxomycetes in a single limited microhabitat. Moreover, plasmodia are almost certainly capable of absorbing some organic material directly from their environment, particularly if it is soluble or finely particulate. In some plasmodia, there may even be extracellular digestion of plant and microbial materials, followed by absorption of the products.

Plasmodia seem incredibly well-adapted to exploit their microhabitats. For example, plasmodia are capable of migrating through extremely small openings, often no larger than a few micrometers. This is to their advantage when penetrating the tiny pores and crevices that exist in substrates such as decaying wood. Aphanoplasmodia appear to require higher levels of moisture in their immediate environment than do phaneroplasmodia, which may account for the fact that only phaneroplasmodia are commonly encountered in nature. Presumably, because of their greater requirement for moisture, aphanoplasmodia spend most of their existence inside a given substrate, emerging to the surface only to sporulate. In contrast, phaneroplasmodia, which can withstand drier conditions, are able to spend more time at the surface of a substrate. It seems likely that this difference is related to the presence of a prominent and well developed slime sheath in the latter. Such a sheath, which would be expected to offer protection against water loss, is lacking in aphanoplasmodia (Alexopoulos 1960).

Substrate Relationships

Although the ability of a plasmodium to migrate some distance from the substrate within which it developed has the potential of obscuring substrate relationships, field observations and collections by numerous workers indicate that the fruiting bodies of particular species of myxomycetes tend to be rather consistently associated with certain types of substrates (Pratt 1934; Gray and Alexopoulos 1968; Stephenson 1988). Some species almost always occur on wood or bark, while others are found more often on dead leaves and other plant debris. Although such associations are not absolute, they are much too frequent to be attributed to chance alone. The reasons for substrate specificity in myxomycetes are not known (Eliasson 1981; Blackwell 1984), but may very well involve the interaction of a number of physical and biotic factors (Stephenson 1988).

Fruiting Season

In temperate regions of the world, the fruiting bodies of myxomycetes first begin to appear with some abundance in early summer and continue to develop until late fall. Although some species may occur during this entire period, provided environmental conditions are favorable, others are seasonal in their fruiting. For example, in eastern North America the fruiting bodies of *Lycogala epidendrum* usually first appear in early summer and may be collected until late fall. *Ceratiomyxa fruticulosa* is common throughout the summer, but is usually absent in the fall. On the other hand, *Metatrichia vesparia* is rarely encountered before late summer (Stephenson 1988). Myxomycetes are not active during the winter in temperate regions, although the fruiting bodies of a few species sometimes appear as a result of a period of unseasonably warm weather. In tropical regions of the world, myxomycetes remain active throughout the entire year, but their fruiting bodies often do not appear in any abundance except at certain times (Lazo 1966). The factors that prompt fruiting are not always as simple as the occurrence of a single unusually favorable period but can be more complex. Rojas et al. (2021) found that the average monthly temperature and accumulated precipitation over the previous four days were the most important climatic variables related to the appearance of fruiting bodies at a study area in Costa Rica.

Temperature and moisture appear to be the primary factors affecting the seasonal distribution of myxomycetes in general (Gray and Alexopoulos 1968). Blackwell and Gilbertson (1984) reported that in moist chamber cultures, *Physarum straminipes* from the Sonoran Desert of the southwestern United States fruited more quickly at 20°C than at 30°C. In an earlier field study carried out in the same region, *P. straminipes* was never collected in the summer; it appeared only during the cooler winter months. On the basis of the data represented by collections of myxomycetes made at regular intervals for one full year from two study areas in Brazil, Maimoni-Rodella and Gottsberger (1980) also concluded that temperature was the single most important factor. They found myxomycetes to be rare when air temperatures were below 14°C, even when moisture was abundant.

Although temperature and moisture are certainly important, other as yet undetermined factors are probably also involved in determining patterns of seasonal distribution for species of myxomycetes with a very limited period of fruiting. Actually, seasonal distribution records based on a considerable number of collections made over several years exist for very few regions of the world. Consequently, our knowledge of this aspect of myxomycete ecology is still rather limited.

Insect Associates

Myxomycetes are known to offer food, shelter, and a breeding place to various species of insects, but the ecological relationships involved are inadequately known. Among the most commonly encountered insect associates of myxomycetes in temperate forests of the Northern Hemisphere are beetles. Members of the family Leiodidae, particularly species of *Anisotoma* and *Agathidium* [Figure 5-1], have been collected repeatedly from myxomycete fruiting bodies and (less frequently) plasmodia. Many of these would seem to be obligate

Figure 5-1. Slime mold beetles. *Agathidium* (left) and *Anisotoma* (left) (©Christoph Benisch).

Figure 5-2. Aethalium of *Reticularia lycoperdon* with beetle entry holes (©Henri Koskinen).

myxomycete specialists and feed only upon fruiting bodies and/or plasmodia of myxomycetes (Wheeler 1984a). In addition to the Leiodidae, representatives of other beetle families such as the the Cerylonidae, Clambidae, Eucinetidae, Lathridiidae, Rhysodidae, Scaphidiidae, and Sphindidae are known to occur in association with myxomycetes (Kylin 2001).

There has been little recognition in either the entomological or ecological literature of these "slime mold" beetles, even though they are rather common. Based on field observations and collections made by Stephenson (unpublished data), it is unusual not to find one or more beetles (either adults or larvae) associated with large fruitings of many of the more common species of myxomycetes. Slime mold beetles are quite small, usually no more than a few millimeters in length, and thus easily overlooked in nature. In fact, their presence is often discovered in a particular fruiting only after the latter has been collected and brought back to the home or laboratory. Beetles burrow into aethalia or pseudoaethalia to feed upon the spore mass. Often, the only evidence of their presence is provided by the "entry holes" they make in the peridium or cortex (Figure 5-2).

In fruitings of *Stemonitis*, beetles tend to occur among the sporangia, where they are hidden from view. When disturbed, members of the genus *Anisotoma* often fall to the substrate and remain motionless. Although capable of flight, these beetles only rarely fly away and thus are fairly easy to capture. Such is not the case for many of the other types of slime mold beetles, which are much more active. It is difficult to assess the effect of the feeding activities of slime mold beetles on myxomycetes. In at least some instances it is probably rather appreciable, since beetles can decimate a large fruiting within a matter of a few days. Beetles that occur on or within the fruiting bodies of myxomycetes are often "dusted" with spores, and it seems likely that these organisms play some role in dispersing the spores of myxomycetes (Blackwell et al. 1982; Blackwell 1984b; Suginji et al. 2019).

Although most collections of slime mold beetles have been made from mature fruiting bodies, plasmodia probably represent a food source that is just as important (Blackwell 1984; Wheeler 1987). If one looks closely, it is not uncommon to find one or more small beetles associated with large plasmodia in nature. This is only one of many aspects of the ecology of slime mold beetles that needs to be investigated more fully. Additional information on slime mold beetles was provided by Russell (1979), Lawrence and Newton (1980), Wheeler (1979, 1984a, 1984b), and Stephenson et al. (1994).

Various species of flies also commonly occur as associates of myxomycetes (Buxton 1954, 2009). Many of these are species that develop from larvae generally regarded as saprophagous or fungivorous (Figure 5-3). At least a few species, however, seem to be restricted to myxomycetes and are not known to breed in any other microhabitat. The flies most commonly encountered as myxomycete associates belong to the family Mycetophilidae, with members of the Sciaridae and Drosophilidae also represented. One of the slime mold flies most likely to be encountered is *Epicypta testata*, which commonly occurs in association with such myxomycetes as *Lycogala epidendrum*, *Reticularia lycoperdon*, *R. splendens*, and *Tubifera ferruginosa*. It may be recognized by the rather distinctive web-like covering that surrounds the puparium in which the adult fly develops. Puparia are located usually within the spore mass of the fruiting body, with only their tips showing. It has been reported (Sellier and Chassain 1976) that the adult fly sometimes deposits its eggs in the plasmodia of *Tubifera ferruginosa* and *Reticularia lycoperdon*. The eggs soon hatch, giving rise to the wormlike larvae. These feed upon the plasmodium until ready to pupate, an event that occurs just as the plasmodium is ready to sporulate. The adult fly, upon emerging from the puparium, comes into contact with the spores present in the mature fruiting body of the myxomycete. Some of these spores adhere to its body and are carried away. In this way the fly helps

Figure 5-3. Slime mold fly feeding on a plasmodium (©Bob Armstrong).

Figure 5-4. Invertebrates commonly associated with the plasmodia or fruiting bodies of myxomycetes. (A) Beetle (©Bob Armstrong). (B) Ant (©Kim Fleming). (C) Snail (©Anne-Marie Rantet Poux). (D) Slug (©Steve Stephenson).

disperse the spores. Except for this one species, the flies associated with myxomycetes have never been studied in any detail. Because existing data on slime mold flies are so very limited, these organisms are even more biologically enigmatic than slime mold beetles.

Various other insects and certain other invertebrates are known to feed upon myxomycetes, including ants, snails, slugs and springtails (Greenslade et al. 2002; Keller and Snell 2002). Both plasmodia and fruiting bodies are subject to being consumed (Figure 5-4). Published reports of larger animals feeding on myxomycetes are exceedingly rare and apparently limited to birds (Sutherland and Crawford 1979; Nikolov and Hristova 2009) feeding upon large plasmodia of what is likely to have been *Fuligo septica*.

Myxomyceticolous Fungi

The fruiting bodies of myxomycetes provide an organic substrate that is colonized by various species of fungi. The majority of these fungi also occur on other types of substrates, but a few species appear to be restricted to myxomycetes and thus are obligately **myxomyceticolous**. Although commonly encountered in nature, the myxomyceticolous fungi have received relatively little attention. Most collectors simply pass over specimens of myxomycetes that show evidence of having been colonized by fungi. Such "moldy" specimens are certainly less attractive; moreover, the presence of the fungus often renders a particular specimen difficult or even impossible to identify.

The fungi that occur on myxomycetes are potentially rather interesting in their own right. The majority of these fungi are hyphomycetes, but some are ascomycetes. Interestingly, many of the species of hyphomycetes are actually the asexual stages of the ascomycetes. *Verticillium rexianum*, a hyphomycete, is the species most likely to be encountered in nature.

The mycelium of this fungus, which occurs as a fuzzy white covering over the surface of the fruiting body, has been recorded from all major groups of myxomycetes and is particularly common on species of *Comatricha* and *Stemonitis*. When the myxomycete involved is a member of the traditional order Physarales, the colonizing fungus is more likely to be *Gliocladium album* or *Sesquicillium microsporum*, two other hyphomycetes.

If there is any evidence of perithecia, then the fungus is probably *Nectriopsis candicans*, an ascomycete. **Perithecia** are flask-shaped fruiting bodies produced by some ascomycetes. In *N. candicans*, they appear as small white mounds, each with a dark spot at the center. *Nectriopsis violacea*, another species in the same genus, has violet or purple perithecia. It has been reported from temperate North America and Europe but does not appear to be particularly abundant in either region. The relatively large number of collections known from southeastern Alaska and Finland suggests that this species may be somewhat more common at higher latitudes (Samuels 1973; Rogerson and Stephenson 1993). Interestingly, this fungus is only known to colonize the aethalia of *Fuligo septica*.

Figure 5-5. Fruiting body of *Arcyria cinerea* colonized by the mycetophilous fungus *Polycephalomyces tomentosus* (©Marius Muessler).

Polycephalomyces tomentosus, a hyphomycete, is another distinctive myxomyceticolous fungus. It occurs almost exclusively on fruiting bodies produced by members of the traditional order Trichiales, including *Arcyria cinerea* (Figure 5-5), *Metatrichia vesparia* and species of *Hemitrichia* and *Trichia*. It is easily recognized by the conspicuous rod-like projections (each consisting of tightly packed fungal hyphae) that radiate from the upper portion of the fruiting body of the myxomycete. More detailed information on myxomyceticolous fungi is provided by Ing (1974, 1976), Ellis and Ellis (1988), Samuels (1973, 1988), and Rogerson and Stephenson (1993).

Bryophyte-Myxomycete Associations

A number of species of myxomycetes are often found in association with bryophytes (mosses and liverworts). It seems likely that many such bryophyte-myxomycete associations are coincidental, since bryophytes and myxomycetes characteristically occur on the same types of substrates, such as decaying wood and litter. Nevertheless, at least a few species of myxomycetes, including *Barbeyella minutissima* and *Lepidoderma tigrinum*, seem to be truly bryophilous, apparently preferring a substrate complex consisting of leafy liverworts growing on decaying coniferous logs. It seems likely that for various other species of myxomycetes the presence of bryophytes on a given substrate provides a more favorable habitat for sporulation than the underlying wood or litter alone. Whether bryophytes also sustain amoeboflagellate cells and plasmodia, the trophic stages of the myxomycete life cycle, is less certain (Stephenson and Studlar 1985). However, there is at least some evidence that this is possible. Stephenson and Rojas (2020) demonstrated that bryophytes are effective spore traps for the spores of myxomycetes, while Stephenson and Stephenson (2022) recorded several species of myxomycetes on samples of bryophyte mats collected in the field and used to prepare moist chamber cultures. Moreover, Schnittler et al. (2010) described a highly specialized assemblage of myxomycetes associated with sandstone ravines in Germany. The assemblage included rarely collected species such as *Diderma ochroceum*.

Myxomycete-Human Relationships

Since most people do not even know that myxomycetes exist, one would not expect these organisms to have much of an impact upon human affairs. It is questionable whether myxomycetes can be said to have any economic value either positive or negative. Although a number of species such as *Diachea leucopodia*, *Fuligo septica*, and *Physarum cinereum* sometimes form extensive fruitings on living plants, including various ornamentals, the grasses of parks and lawns, and agricultural crops, very little, if any, damage is done to the plants involved, and the fruiting bodies are washed away with the first rain.

The sight of a large and conspicuous plasmodium can sometimes excite people who are not familiar with myxomycetes. An unusually spectacular mass appearance of the plasmodia of *Fuligo septica* in a suburb of Dallas, Texas, even made the news nationwide in the United States in 1973. The plasmodia of *F. septica* are usually bright yellow and often spread over large areas. The sudden, unexplained appearance of these "pulsating yellow blobs" in lawns and other similar situations caused a virtual panic among the some of the local inhabitants. The blobs broke apart when blasted with water, but the pieces continued to crawl about and increase in size. To some people, this seemed to suggest the possibility that they were indestructible aliens from outer space, or at the very least, menacing mutant bacteria that could take over the earth. Eventually, the blobs were identified as nothing more than a stage in the life cycle of an organism unfamiliar to most people, and the excitement came to an end (Alexopoulos and Mims 1979; Sharnoff 1991).

Although one would hardly look upon myxomycetes as likely items of food, at least two species have been reported to be consumed by humans. In the state of Veracruz in Mexico, both the plasmodium of *Fuligo septica* and the very early developing aethalia of *Reticularia lycoperdon* are collected, fried, and eaten by some of the Indigenous Peoples. The local name used to refer to these edible myxomycetes is *caca de luna* (López et al. 1982; Villarreal 1983).

Taxonomy and Classification of the Myxomycetes

About 1200 species of myxomycetes have been described to date (Lado 2005–2022). The traditional classification used for the group placed these in six different taxonomic orders—the Ceratiomyxales, Echinosteliales, Liceales, Physarales, Stemonitales, and Trichiales. This system of classification, which was based upon spore color and easily discernible morphological features, has been used in virtually all previous monographs of the myxomycetes, including the first edition of this book. However, it does not properly reflect evolutionary relationships within the group, as revealed by studies, particularly those involving molecular biology, carried out over the past 25 years. For example, the traditional order Ceratiomyxales (which consists of the single genus *Ceratiomyxa* with four species) is now known to be distinctly different from the other orders and is now assigned to a totally different taxonomic class, the Ceratiomyxomycetes. Based on a phylogeny constructed from 18S rDNA sequences (e.g., Fiore-Donno et al. 2005, 2008, 2010, 2013), myxomycetes in the other orders make up a monophyletic group which splits into two basal groupings (or clades). A **clade** is a grouping of taxa made on the basis of the presumed shared evolutionary history rather than on shared morphological features. The latter is the basis for the traditional classification system used for myxomycetes in all but the most recent treatments of these organisms. However, relationships among taxa within a clade remain subject to change, as recently demonstrated for the traditional order Physarales (García-Martín et al. 2023).

The first clade (the subclass Columellomycetidae) consists mostly of members of the traditional orders Stemonitales and Physarales, which have dark (different tints of brown to black) spores, along with members of the traditional order Echinosteliales. The second clade (the subclass Lucisporomycetidae) includes the traditional orders Liceales and Trichiales with brightly colored spores (including red, orange, purple, yellow, and olive). However, these traditional orders no longer hold together in the manner in which they have been circumscribed in the past. Instead, a number of superorders are recognized in each of the two subclasses, with each made up of several orders. For example, the subclass Lucisporomycetidae contains four orders, none of which corresponds exactly to any of the traditional orders, while the subclass Columellomycetidae contains five different newly recognized orders. The classification system used herein essentially follows that outlined by Leontyev and Schnittler (2017) and Leontyev et al. (2019), but in order to simplify the identification keys, some of the morphological features used to delimit taxa in the more traditional treatments of the myxomycetes have been retained.

The fact that the vast majority of myxomycetes can be assigned to one of two groups on the basis of the spore color in mass is not a new concept. It was first proposed by Joszef Rostafinski (1875) and then used in Arthur Lister's *A Monograph of the Mycetozoa* (1894), which was the standard reference on the myxomycetes at the end of the 19th and during the early part of the 20th century. This classic monograph first appeared in 1894, but revised and expanded versions were published in 1911 and 1925 by his daughter Gulielma Lister (Lister 1911, 1925).

It should be noted that the concept of species used herein is that of the morphological species (or **morphospecies**) in which species are distinguished solely on the basis of morphological features (Walker and Stephenson 2016), which may or may not reflect evolutionary relationships. Sometimes, a single morphospecies may consist of two or more taxonomic entities, which can be revealed only after detailed investigations of their morphology and comparisons of their DNA sequences. If these taxonomic entities are considered to be distinct enough, they can be described as species new to science. For example, Leontyev et al. (2015) reported that *Tubifera ferruginosa*, one of the most recognizable of all myxomycetes, actually consists of a complex of at least seven species. An even larger complex is represented by *Lycogala epidendrum*, which may encompass more than 60 species (Leontyev et al. 2023).

CHAPTER 6
Classification and Identification

Taxonomic Names

Unlike many of the more familiar organisms found in nature, myxomycetes do not have common names. As a result, the only name available for use is the taxonomic name. The taxonomic name applied to a particular species of myxomycete is a binomial and consists of a unique combination of two names—a genus name and a specific epithet. For example, *Lycogala* is the genus name and *epidendrum* the specific epithet for one of the more conspicuous and widely distributed myxomycetes, *Lycogala epidendrum*.

The binomial that has been given to a particular species of myxomycete is complete only when it is followed by what is known as the authority, the last name (or an abbreviation of the name) of the person (or persons) who first described the species in question. Since taxonomic concepts are subject to change, a species can sometimes be transferred to a genus other than the one to which it was assigned when originally described. When this happens, the name of the person who formally proposed the change is added to the authority and the original name is placed in parentheses. For example, *Lycogala epidendrum* (L.) Fr. indicates that Carl Linnaeus (name abbreviated as "L.") was the first to apply a name to this very common myxomycete. Linnaeus considered it to be a small puffball and placed the species in the mushroom genus *Lycoperdon*. Later, Elias Magnus Fries (name abbreviated as "Fr.") proposed that the species be reassigned to the myxomycete genus *Lycogala*, a concept still followed.

Genus and species are but two levels in a hierarchical system of classification in which organisms are grouped on the basis of common characteristics. Two or more species that share many of the same characteristics are placed in the same genus (plural: genera), while genera with common characteristics are grouped in the same family. For example, *Lycogala* is a member of the family Reticulariaceae, which also includes such genera as *Reticularia* and *Tubifera*. Likewise, families with common characteristics are grouped in the same order. Correspondingly, orders that share the same characteristics are grouped in the same class, classes in the same phylum (plural: phyla), and phyla in the same kingdom. Sometimes, additional levels of classification such as superorders and subclasses are recognized. This is the case for the most recent taxonomic treatment of the myxomycetes.

As used in this book, the myxomycetes are considered to represent a single phylum (the Eumycetozoa), within which two classes are recognized. The first class (the Ceratiomyxomycetidae) is very small and consists of just one order (the Ceratiomyxales) containing three genera. Members of only one genus (*Ceratiomyxa*) produce macroscopic fruiting bodies and are considered as myxomycetes.

Both the general structure and the life cycle of members of the genus *Ceratiomyxa* differ from what is considered typical for myxomycetes, and the relationship of *Ceratiomyxa* to other myxomycetes has long been questioned. The most important distinguishing feature of *Ceratiomyxa* is that each spore is produced on an individual, threadlike stalk. As such, the spores are not enclosed within a fruiting body as is the case for all other myxomycetes. It has been hypothesized that the spore and the stalk on which it is borne should be regarded as the equivalent of a miniature fruiting body, which would make the "fruiting body" of *Ceratiomyxa* a composite structure unlike anything else found in the myxomycetes. This composite structure, which typically consists of a series of erect, simple or branched columns, produces an almost incalculable number of spores. The spores of *Ceratiomyxa* differ from those of other myxomycetes in one other respect. When a spore germinates, it gives rise to a single, spherical protoplast that elongates to form a distinctive thread stage that is quite unlike anything found in the life cycle of a typical myxomycete. This structure then rounds up and undergoes cleavage to produce four and then eight cells. Finally, these cells separate and become transformed into swarm cells. The significance of these differences is not known.

The second class (Myxomycetes) includes all the other organisms recognized as "true" myxomycetes. The dichotomous keys below consist of a series of numbered couplets, with each couplet made up of a pair of statements (or leads) composed of one or more contrasting characters. For the most part, the characters are features, such as color of the spore mass, or structures, such as the type of fruiting body, that can be observed by the naked eye or with the use of a good hand lens. Because most myxomycetes are very small, however, some characters can be determined only with the aid of a microscope. Each lead is followed by a number, which indicates the couplet to use next or the name of a species of myxomycete. To identify a particular specimen, one should begin with the first couplet, the pair of leads beginning with "1," and select the lead that applies to the specimen being identified. This same procedure is repeated for each successive couplet until the order and ultimately the species to which the myxomycete belongs is reached. When a species is indicated, the identification of a myxomycete is then confirmed or rejected by checking it against the illustration and description provided for each species included in the keys.

Although it should be possible to identify the majority of myxomycetes likely to be encountered with these keys, the keys are not comprehensive and do not include species that are rare or infrequently collected. Any myxomycete that does not "key out" in the keys often can be identified from the information on similar or look-alike species provided in the "Comments" section for each of the 58 primary species or, if necessary, with the aid of one of the more technical taxonomic monographs of the group (e.g., Martin and Alexopoulos 1969, Ing 1999, Poulain et al. 2011, Stephenson 2021).

Keys to Common Species of Myxomycetes

Key to the classes of the phylum Eumycetozoa

"Fruiting body" a composite structure typically consisting of a series of erect, simple or branched, usually white but sometimes pink or pale yellow columns or a pitted cushion-like mass; spores attached individually by threadlike stalks...**Class Ceratiomyxomycetes** (page 55)

1. Fruiting body clearly not as above; spores occurring in a mass within the fruiting body and enclosed (at least during the early stages of development) by a peridium...**Class Myxomycetes** (page 55)

Class Ceratiomyxomycetes

The class Ceratiomyxomycetes contains only the single order Ceratiomyxales. As already indicated, the members of the Ceratiomyxales are clearly different from other myxomycetes, and based on molecular data they should be considered as a sister group to the "true" myxomycetes. Only one genus (*Ceratiomyxa*) of the three assigned to the class Ceratiomyxomycetes shows any resemblance to the "true" myxomycetes. The genus *Ceratiomyxa* contains ten species, but only one of these (*C. fruticulosa*) is likely to be encountered by the typical collector. However, this species (*C. fruticulosa*) is one of the most common and widely distributed myxomycetes in the world. Two other species, much less common, are restricted to subtropical or tropical regions.

Figure 6-1. *Ceratiomyxa fruticulosa* (poroid form) (©Kim Fleming)

Figure 6-2. *Ceratiomyxa fruticulosa* (filamentous form) (©Kim Fleming)

Class Myxomycetes

The class Myxomycetes contains all of the organisms recognized as "true" myxomycetes. The latter produce fruiting bodies that display a wide range in color, size, and structure. Some are among the most beautiful miniature structures found in nature. The class Myxomycetes is made up of two subclasses, with the primary distinguishing feature being the color of the spores in mass.

Key to the subclasses of the class Myxomycetes

1. Spores in mass brightly colored (typically yellow, orange or red but sometimes brown, violet or olivaceous); columella lacking; capillitium, if present, tubular and nearly smooth to highly ornamented .... **Subclass Lucisporomycetidae** (below)

1. Spores in mass typically brown to black; columella often present; capillitium usually present, solid, often branched and anastomosed, attached to the columella .... **Subclass Columellomycetidae** (page 62)

Subclass Lucisporomycetidae

This subclass is made up of two superorders that essentially include some of the myxomycetes placed in the traditional orders Liceales and Trichiales.

Key to the superorders of the class Lucisporomycetidae

Peridium perforated or more rarely entire; capillitium absent....... Superorder Cribrariidia (below)

1. Peridium not perforated, capillitium present or absent .... Superorder Trichiidia (page 57)

Superorder Cribrariidia

This superorder contains one order (the Cribrariales) and the two genera most likely to be encountered by collectors. Members of the order have two distinguishing features. The first of these is a netlike covering over the spore mass of the fruiting body. In some species the netlike covering has obvious thickened areas called **nodes** (Figure 6-3), but in other species nodes are lacking (Figure 6-4). This distinctive covering is composed of portions of the peridium that persist in mature fruiting bodies. The second feature is the presence of minute, granular structures called **dictydine granules** in the fruiting bodies. In addition, with only a few exceptions such as *Lindbladia tubulina*, all of the species in this family produce stalked sporangia as fruiting bodies.

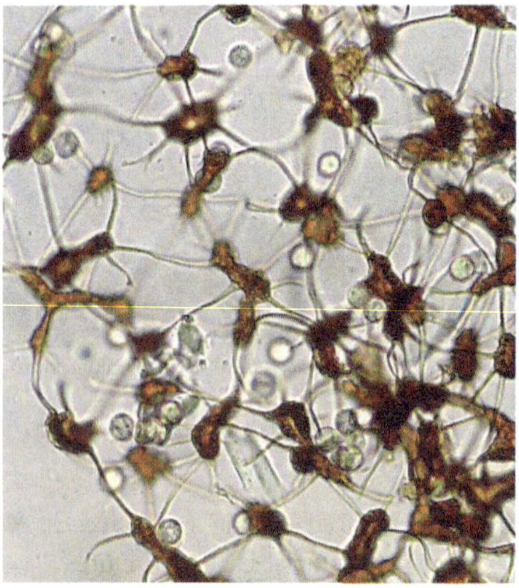

Figure 6-3. Peridial net of *Cribraria intricata* in which nodes are present (100X) ©Alain Michaud).

Figure 6–4. Peridial net of *Cribraria rufa* in which nodes are not present (630X) (©Alain Michaud).

Key to common species in the superorder Cribrariidia

1. Peridial net consisting of longitudinal ribs connected by delicate transverse filaments .... ***Cribraria cancellata*** (page 89)

1. Peridial net consisting of a network of filaments that connect and radiate from expanded or thickened portions (nodes)...2

2. Sporangium short-stalked or nearly sessile, usually in tight clusters, peridial net weakly developed and without clearly differentiated nodes.......................... ***Cribraria argillacea*** (page 88)

2. Sporangium with a relatively long stalk, peridial net well developed with clearly differentiated nodes...3

3. Stalks very long and slender, usually at least six times as long as the diameter of the sporotheca .... ***Cribraria microcarpa*** (page 92)

3. Stalk shorter, less than six times as long as the diameter of the sporotheca .... ***Cribraria intricata*** (page 90)

Figure 6-5. *Cribraria cancellata* (©Yuri Novozhilov)

Figure 6-6. *Cribraria argillacea* (©Alain Michaud)

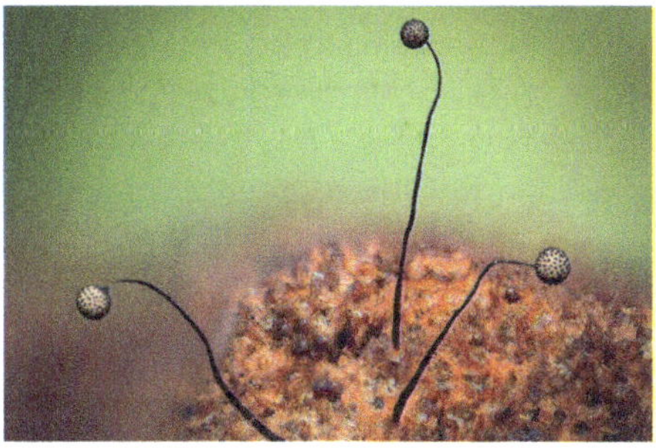
Figure 6-7. *Cribraria microcarpa* (©Marius Muessler)

Figure 6-8 *Cribraria intricata* (©Alain Michaud)

Superorder Trichiidia

The superorder Trichiidia contains three orders, one of which (the Trichiales) corresponds to the traditional order of the same name. The two other orders (Liceales and Reticulariales) are included within the traditional concept of the Liceales, which now also includes the Cribrariidia.

Key to the orders of the superorder Trichiidia

Capillitium present (or very rarely absent in a few species), tubular and usually very highly ornamented .... **Order Trichiales** (below)

1. Capillitium usually absent but if present attached to the peridium with funnel-like junctions .. 2

Capillitium absent, spores warted to nearly smooth .... **Order Liceales** (page 60)

2. Capillitium usually absent but when present attached to the peridium with funnel-like junctions, spores reticulate .... **Order Reticulariales** (page 61)

Order Trichiales

Members of the order Trichiales are usually fairly easy to recognize, even in the field. The spore mass is typically more or less brightly colored (rarely gray or light brown), a characteristic shared with few other myxomycetes. Most species in the order produce stalked or sessile sporangia, but a few produce plasmodiocarps. A capillitium consisting of solid or tubular, smooth or sculptured, free or attached threadlike elements is always present, but in some instances it may be so limited in extent as to be easily overlooked. As traditionally circumscribed, this order contains two families—the Dianemaceae and the Trichiaceae. Species assigned to the Dianemaceae are characterized by a capillitium in which the individual elements are threadlike and nearly or entirely solid, while species assigned to the Trichiaceae have a capillitium in which the individual elements are tubular. Virtually all the commonly encountered representatives of this order, including species in such genera as *Arcyria*, *Hemitrichia*, *Metatrichia*, *Perichaena*, and *Trichia*, are members of the second of these two families.

Key to common species in the order Trichiales

1. Fruiting body a plasmodiocarp .... ***Hemitrichia serpula*** (page 108)

1. Fruiting body a stalked or sessile sporangium 2

2. Sporangia sessile or (occasionally) very short-stalked 3

2. Sporangia with a well-developed stalk 6

3. Sporangia flattened, often angular, red-brown to purple-brown or almost black; bright yellow inside .... ***Perichaena depressa*** (page 123)

3. Sporangia not noticeably flattened, rounded, color not as above 4

4. Sporangia subglobose to globose; spore mass orange-brown to dull brown, peridium iridescent .... ***Prototrichia metallica*** (page 134)

4. Sporangia pulvinate, cylindrical, ovoid, or somewhat elongated; spore mass yellow, orange-yellow or yellow-brown 5

5. Sporangia bright yellow, clavate, shining when fresh; spores distinctly reticulate .... ***Trichia favoginea*** (page 142)

5. Sporangia dull yellow-brown or olivaceous, sometimes with a short dark stalk, usually not shining; spores minutely warted .... ***Trichia varia*** (page 144)

6. Peridium with conspicuous wart-like structures present .... ***Hemitrichia pardina*** (page 107)

6. Peridium with no evidence of wart-like structures present 7

Figure 6-9. *Hemitrichia serpula* (©Jonathan Mays)

Figure 6-10. *Perichaena depressa* (©Laurie Leonard)

Figure 6-11. *Prototrichia metallica* (©John Shadwick)

Figure 6-12. *Trichia favoginea* (©Kim Fleming)

Figure 6-13. *Trichia varia* (©Alain Michard)

Figure 6-14. *Hemitrichia pardina* (©Alison Pollack)

Figure 6-15. *Metatrichia vesparia* (©Yuri Novozhilov)

6-16. *Arcyria denudata* (©Alain Michaud)

Figure 6-17. *Arcyria cinerea* (©Henri Koskinen)

Figure 6-18. *Hemitrichia calyculata* (©Jeffery Gruber)

7. Sporangia red, maroon or red-brown.......... 8
7. Sporangia whitish or pale gray to yellow to yellow-brown.......... 9
8. Sporotheca with a distinct, preformed operculum, often firmly united into clusters of several to many.......... ***Metatrichia vesparia*** (page 121)
8. Sporotheca without a preformed operculum, typically scattered to gregarious.......... ***Arcyria denudata*** (page 77)
9. Sporangia typically white or pale gray; peridium persisting as an inconspicuous shallow calyculus.......... ***Arcyria cinerea*** (page 76)
9. Sporangia yellow or yellow-brown; peridium persisting as a fairly conspicuous, sometimes relatively deep, calyculus 10
10. Sporangia typically scattered to gregarious; stalk well-developed; peridium persisting as a relatively shallow but conspicuous calyculus.......... ***Hemitrichia calyculata*** (page 106)
10. Sporangia typically crowded, short-stalked, peridium persisting as a deep calyculus 11

Figure 6-19. *Arcyria versicolor* (©Alain Michaud)

Figure 6-20. *Trichia decipiens* (©Kim Fleming)

11. Capillitium strongly elastic, netted, tending to remain together as a single unit .... ***Arcyria versicolor*** (page 78)
11. Capillitium not noticeably elastic, consisting of free elaters, tending to break apart .... ***Trichia decipiens*** (page 141)

Order Liceales

This order contains those myxomycetes with very small, mostly sessile fruiting bodies in which a capillitium is lacking and the spore mass is enclosed by a thick, coarse, dark peridium. The vast majority of the species in the Liceales are currently accommodated in the genus *Licea*. A second genus (*Listerella*), with a single species, differs from the other members of the order by having a capillitium; it is not included in this book. The fruiting bodies (mostly sporangia) of members of the Liceales are so small that they usually go undetected in the field, although several species commonly appear on bark and other substrates when these are placed in moist chamber cultures.

Key to common species in the genus *Licea*

1. Sporangium elongated, dehiscence by a single longitudinal fissure .... ***Licea biforis*** (page 117)
1. Sporangium not elongated, dehiscence not as above 2
2. Sporangium angular, sessile, with prominent lines or ridges, dehiscence not as above .... ***Licea minima*** (page 119)

Figure 6-21. *Licea biforis* (©Steve Young)

Figure 6-22. *Licea minima* (©Alain Michaud)

Figure 6-23. *Licea operculata* (©Alison Pollack)

Figure 6-24. *Licea kleistobolus* (©Alison Pollack)

2. Sporangium pulvinate or subglobose; stalked or sessile, dehiscence by means of a delicate preformed operculum..3
3. Sporangium stalked...***Licea operculata*** (page 119)
3. Sporangium sessile ....***Licea kleistobolus*** (page 117)

Order Reticulariales

In sharp contrast to the situation that exists in the Liceales, the fruiting bodies (usually either aethalia or pseudoaethalia) produced by members of the Reticulariales are often quite large and thus are far more likely to be noticed in nature. As a result, certain common species in such genera as *Lycogala*, *Reticularia*, and *Tubifera* are among the best known myxomycetes. Although included in the key given below, recent molecular evidence (Leontyev et al. 2019) indicates that the genus *Dictydiaethalium* is more properly assigned to the family Dianemataceae. However, for convenience, the position of this genus in the traditional classification used for myxomycetes is retained in this book. Also, in the key given below, the fruiting body of *Lycogala* is considered to be an aethalium, which is the traditional concept used for the members of this genus. However, it is probably best regarded as an unusually large sporangium with a true capillitium and not a pseudocapillitium present.

Key to common species in the order Reticulariales

1. Fruiting body a pseudoaethalium...2
1. Fruiting body an aethalium...3
2. Pseudoaethalium flat, consisting of individual sporangium-like units compacted into a palaside-like layer ....***Dictydiaethalium plumbeum*** (page 94)
2. Pseudoaethalium consisting of numerous sessile sporangia crowed together, the individual units cylindrical, pseduocapillitium absent ....***Tubifera ferruginosa*** (page 116)
3. Aethalium pulvinate, pseudocapallitium consisting of perforated plates or irregular threadlike elements ....***Reticularia splendens*** (page 115)
3. Aethalium globose to pulvinate, capillitium consisting of tubular elements ....***Lycogala epidendrum*** (page 120)

Figure 6-25. *Dictydiaethalium plumbeum* (©Kim Fleming)

Figure 6-26. *Tubifera ferruginosa* (©Kim Fleming)

Figure 6-27. *Reticularia splendens* (©Richard Orr)

Subclass Columellomycetidae

Members of this subclass, as the name suggests, are characterized by the presence of a columella, but this structure is sometimes absent and other features assume greater importance for identification. Another name for the subclass, the dark-spored myxomycetes, reflects the brown to black color of the spores in mass. However, the members of the genus *Echinostelium* (in the superorder Echinosteliidia) have spores that are essentially colorless. Members of the second superorder (the Stemonitidia) have dark spores, usually dark brown or black; the same color is also typical for the second genus (*Barbeyella*) in the Echinosteliidia. This superorder encompasses the members of the traditional orders Stemonitales and Physarales.

Key to the superorders of the subclass Columellomycetidae

1. Spores essentially colorless (may appear slightly pinkish or yellowish); capillitium, if present, scanty and poorly developed, stalk granular.......................... **Superorder Echinosteliidia** (page 63)

1. Spores usually dark brown or black; capillitium always present and usually abundant; stalk granular, fibrous or amorphous... **Superorder Stemonitidia** (page 63)

Superorder Echinosteliidia

This superorder contains the single order Echinosteliales, which does not include all of the taxa assigned to the traditional order of the same name. With the exception of *Barbeyella minutissima* (page 80) all members of the Echinosteliales have colorless spores. There are more than a dozen species of *Echinostelium*, but all are small and virtually impossible to detect in the field. Only the single species *Echinostelium minutum* (page 103) is illustrated and described in this book.

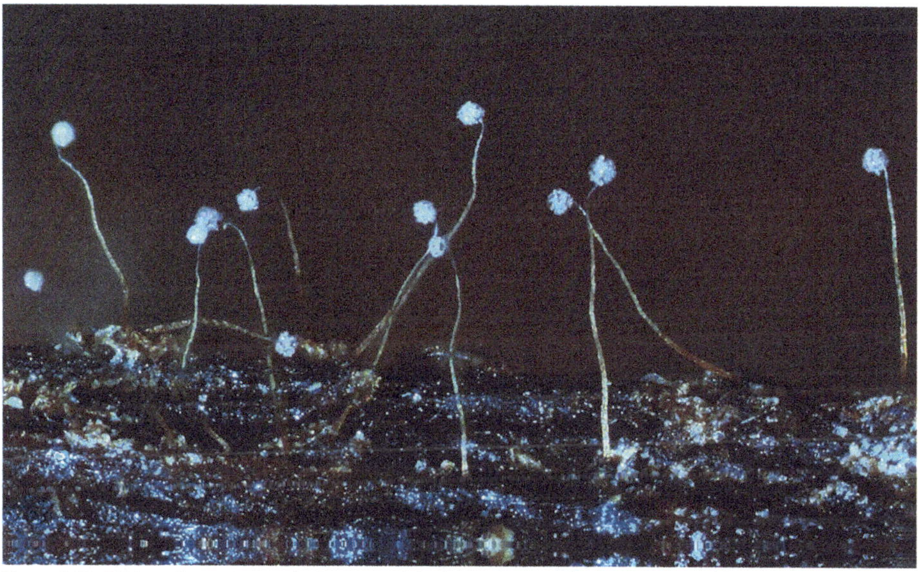

Figure 6-28. *Echinostelium minutum* (©Yuri Novozhilov)

Superorder Stemonitidia

There are four orders in the superorder Stemonitidia. One of these (the Clastodermataceae) is assigned to the Echinosteliales in traditional taxonomic treatments of the myxomycetes.

Key to the orders of the Stemonitidia

1. Stalk with a basal portion filled with granular material.........**Order Clastodermatales** (page 64)

1. Stalk not as above 2

2. Capillitium with funnel-shaped ends which are attached to the peridium..**Order Meridermatales** (page 64)

2. Capillitium not as above; sometimes attached to the peridium but with no evidence of funnel-shaped ends 3

3. Stalk fibrous or amorphous, columella almost always present and usually extending to near the apex of the sporotheca .... **Order Stemonitidales** (page 72)

3. Stalk filled with amorphous material, columella present or absent, when present not usually extending to near the apex of the sporotheca .... **Order Physarales** (page 65)

Order Clastodermatales

In the traditional classification of the myxomycetes, members of this order were placed in the order Echinosteliales. However, they differ in having brown and not essentially colorless spores. The order Clastodermatales contains a single family (Clastodermataceae), one genus, and three described species. Only one of these (*Clastoderma debaryanum*) is relatively common. It is easy to recognize by having a stalk divided into two portions separated by a prominent oval swelling (Figure 6-29).

Figure 6-29. *Clastoderma debaryanum* (©Steve Young)

Order Meridermatales

This order consists of a single family (Meridermataceae) and one genus (*Meriderma*). The species that make up this genus were not recognized as distinct until 2011. Until then, they were placed in the genus *Lamproderma*, which they closely resemble with respect to overall morphology. The critical difference that distinguishes members of the two genera is the presence in *Meriderma* of a capillitium with funnel-shaped ends (Figure 6-32) that are attached to the peridium. The two species of *Meriderma* most likely to be encountered are *Meriderma carestiae* (Figure 6-30) and *Meriderma cribrarioides* (Figure 6-31).

Figure 6-30. *Meriderma carestiae* (©Alain Michaud)

Figure 6-31. *Meriderma cribrarioides* (©Alain Michaud)

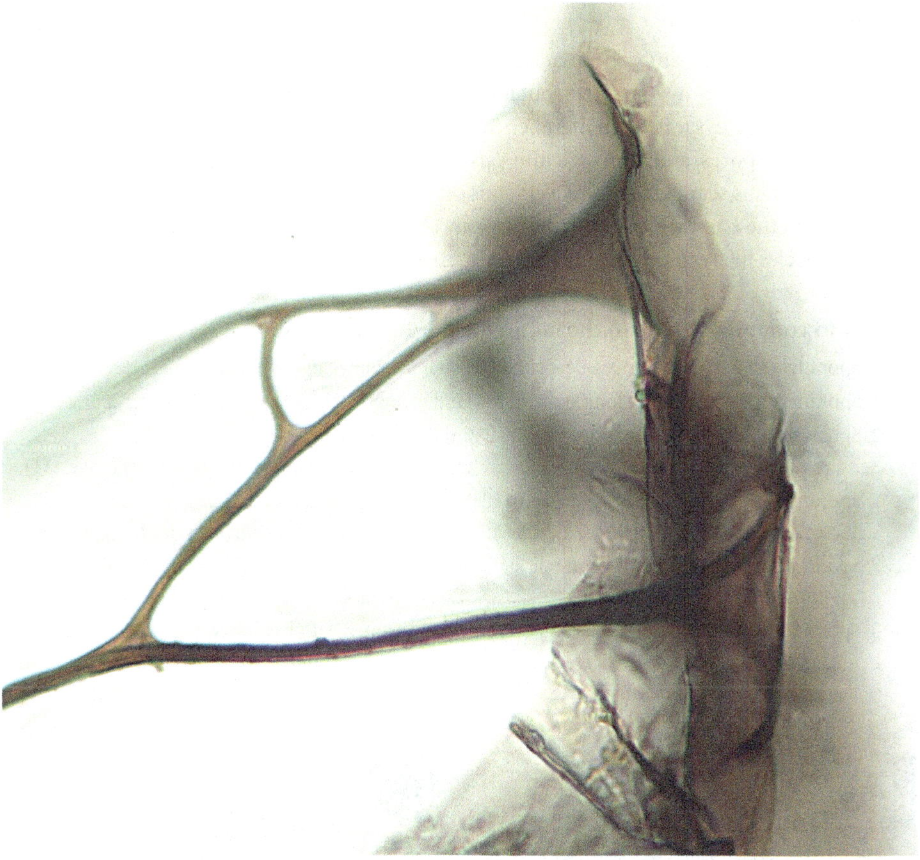

Figure 6-32. Funnel-shaped capillitial tips of *Meriderma spinulosporum* (630X) (©Alain Michaud).

Order Physarales

The order Physarales as currently circumscribed contains three families. Two of these, the Physaraceae and the Didymiaceae, essentially make up the traditional order of the same name. However, the third family (Lamprodermataceae) is included in the traditional order Stemonitales.

Key to the families of the order Physarales

1. Lime lacking from all parts of the fruiting bod .... **Family Lamprodermataceae** (page 66)

1. Lime usually present in some part of the fruiting body 2

2. Lime deposits present on the peridium and the capillitium, the latter often in the form of lime nodes .... **Family Physaraceae** (page 70)

2. Lime deposits almost always absent from the capillitium; if present not in the form of lime nodes .... **Family Didymiaceae** (page 68)

Lamprodermataceae

As noted above, the Lamprodermataceae is included in the Stemonitales in the traditional classification used for myxomycetes, but molecular evidence indicates that it is better placed in a reconfigured Physarales. All members of this family of myxomycetes typically have dark brown to black spores, and lime (calcium carbonate) is lacking from all parts of the fruiting body, or present as scanty needle-like crystals on the peridium. Six different genera are recognized in the Lamprodermataceae, but only one of these (the type genus *Lamproderma*) contains species likely to be encountered on a fairly regular basis.

Key to common species in the family Lamprodermataceae

1. Sporangia short-stalked or sessile, primarily in alpine snowbank habitats 2
1. Sporangia with a relatively long stalk, in lowland habitats 3
2. Sporangia up to 2 mm tall; sporotheca 1–2 mm in diameter .... ***Lamproderma sauteri*** (page 111)
2. Sporangia to 1.5 mm tall, sporotheca 0.4–0.9 mm in diameter .... ***Lamproderma arcyrioides*** (page 109)
3. Sporangia usually less than 2 mm tall, stalk slender, not typically associated with bryophytes .... ***Lamproderma scintillans*** (page 112)
3. Sporangia than 2 mm tall, stalk robust, typically associated with bryophytes .... ***Lamproderma columbinum*** (page 110)

Figure 6-33. *Lamproderma sauteri* (©Alain Michaud)

Figure 6-34. *Lamproderma arcyrioides* (©Alain Michaud)

Figure 6-35. *Lamproderma scintillans* (©Joe Takano)

Figure 6-36. *Lamproderma columbinum* (©Alain Michaud)

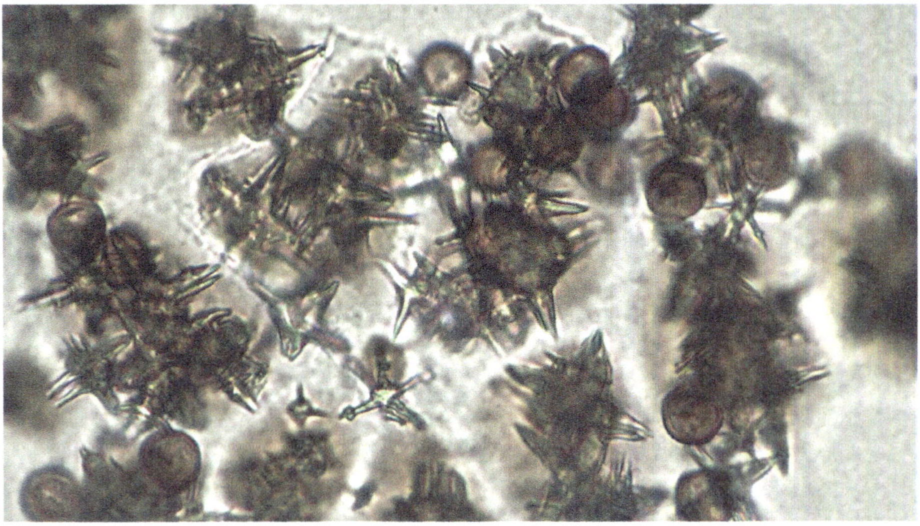

Figure 6-37. Crystalline lime of *Didymium nigripes* (430x) (©Alain Michaud).

Figure 6-38. Limeless capillitium of *Didymium dubium* (630X) (©Alain Michaud)

The families Physaraceae and Didymiaceae

The single most important characteristic of the two families that make up the traditional order Physarales or two of the three families recognized in the more modern concept of the order is the presence of lime (calcium carbonate) deposits, which may occur in the peridium, capillitium, or stalk of the fruiting body. The presence of lime is usually an obvious feature, but under certain environmental conditions fruiting bodies are sometimes produced that have very little lime. Such fruitings are often difficult to identify. The peridium in certain species of *Perichaena* (*P. depressa* is one example) sometimes appears limy due to deposits of calcium oxalate (Schoknecht and Keller 1977). *Perichaena*, however, as a member of the Trichiales, has brightly colored spores and thus is unlikely to be misidentified as a member of the Physaraceae or Didymiaceae, in which the spore mass ranges from deep violaceous brown to purple-brown to black in color. Members of the family Didymiaceae differ in one important respect from members of the family Physaraceae. In the Physaraceae the capillitium contains lime (Figure 2-13), while in the Didymiaceae lime is absent from the capillitium (Figure 6-38). In addition, species belonging to three genera (*Didymium*, *Lepidoderma*, and *Mucilago*) assigned to the Didymiaceae are characterized by the presence of crystalline lime (Figure 6-37) in the peridium. Other members of the Physaraceae have amorphous (usually granular) deposits of lime.

Key to common species in the family Didymiaceae

1. Fruiting body an aethalium..***Mucilago crustacea*** (page 122)
1. Fruiting body a sporangium 2
2. Lime absent from the peridium...***Diachea leucopodia*** (page 93)
2. Lime present on the peridium 3
3. Lime present on the peridium in the form of distinct plates or as small crystals..........................4
3. Lime present on the peridium in the form of a continuous layer...6
4. Lime present on the peridium in the form of distinct plates..
....***Lepidoderma tigrinum*** (page 115)
4. Lime present on the peridium in the form of crystals, these forming a powdery layer................5
5. Sporotheca hemispherical; stalks usually no more than half the total height of the fruiting body
....***Didymium melanospermum*** (page 102)

Figure 6-39. *Mucilago crustacea* (©Gary Laursen)

Figure 6-40. *Diachea leucopodia* (©Alain Michaud)

Figure 6-41. *Lepidoderma tigrinum* (©Marius Muessler)

Figure 6-42. *Didymium melanospermum* (©Yuri Novozhilov)

5. Sporotheca globose; stalks usually two-thirds or more of the total height.. ....*Didymium iridis* (page 101)
6. Fruiting body stalked..*Diderma floriforme* (page 97)
6. Fruiting body sessile...7
7. Fruiting body a plasmodiocarp ....*Diderma effusum* (page 95)
7. Fruiting body a sporangium 8
8. Fruiting body pulvinate, peridium pink.. *Diderma testaceum* (page 100)
8. Fruiting body more or less globose, peridium white..........................*Diderma niveum* (page 99)

Figure 6-43. *Didymium iridis* (©Joe Takano)

Figure 6-44. *Diderma floriforme* (©Alain Michaud)

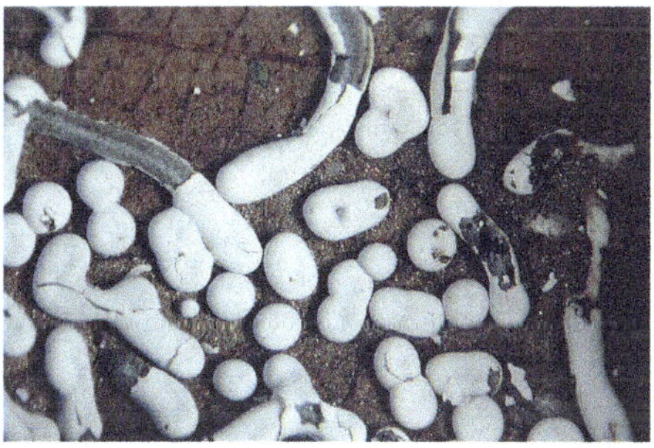

Figure 6-45. *Diderma effusum* (©Laurie Leonard)

Figure 6-46. *Diderma testaceum* (©Alain Michaud)

Figure 6-47. *Diderma niveum* (©Alain Michaud)

Family Physaraceae

Physarum, with more than one hundred described species, is the largest genus in the family Physaraceae and thus among the myxomycetes most likely to be encountered in nature. Fruiting bodies produced by members of the Physaraceae are most often sporangia, but some species produce plasmodiocarps or (more rarely) aethalia. Other common genera include *Badhamia*, *Craterium*, *Diachea*, and *Fuligo*.

Key to common species in the family Physaraceae

1. Fruiting body an aethalium .... ***Fuligo septica*** (page 105)
1. Fruiting body a sporangium 2
2. Fruiting body stalked (sometimes with a weak stalk that is essentially an extension of the hypothallus but never erect) 5
2. Fruiting body clearly sessile with no evidence of a stalk 3
3. Fruiting body fattened from side to side .... ***Physarum bivalve*** (page 127)
3. Fruiting body globose or nearly so 4
4. Peridium yellow, fruiting bodies heaped .... ***Physarum virescens*** (page 133)
4. Peridium white or gray, fruiting bodies gregarious to scattered .... ***Physarum cinereum*** (page 128)

Figure 6-48. *Fuligo septica* (©Steve Stephenson)

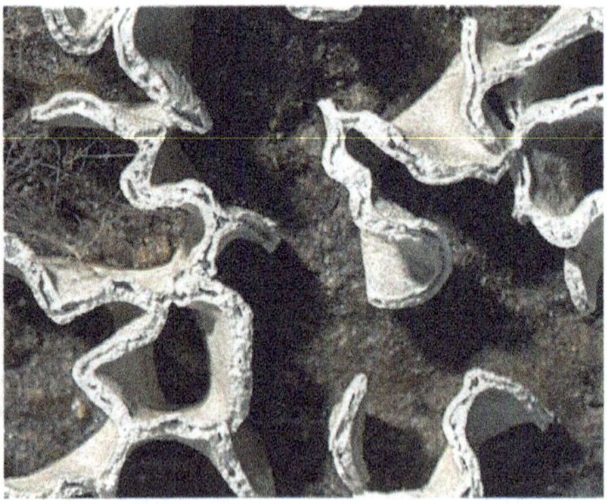

Figure 6-49. *Physarum bivalve* (©Alain Michaud)

Figure 6-50. *Physarum virescens* (©Raymond McNeil)

Figure 6-51. *Physarum cinereum* (©Yuri Novozhilov)

5. Stalk weak, often not clearly differentiated..6
5. Stalk sometimes short but clearly differentiated...8
6. Peridium smooth and shiny, tough...***Leocarpus fragilis*** (page 114)
6. Peridium rough, thin and fragile..7
7. Peridium yellow...***Physarum albescens*** (page 125)
7. Peridium white..***Badhamia utricularis*** (page 79)
8. Sporotheca goblet-shaped to subcylindrical, dehiscence circumscissle, often by a preformed operculum, lower portion of peridium persisting as a deep cup ....
....***Craterium minutum*** (page 86)
8. Sporotheca variously shaped, dehiscence never circumscissle 9

Figure 6-52. *Leocarpus fragilis* (©Henri Koskinen)

Figure 6-53. *Physarum albescens* (©Yuri Novozhilov)

Figure 6-54. *Badhamia utricularis* (©Henri Koskinen)

Figure 6-55. *Craterium minutum* (©Alain Michaud)

Figure 6-56. *Physarella oblonga* (©Todd Elliott)

Figure 6-57. *Physarum polycephalum* (©Mark Bower)

Figure 6-58. *Physarum album* (©Alain Michaud)

Figure 6-59. *Physarum globuliferum* (©Joe Takano)

9. Sporotheca thimble-shaped; peridium splitting into reflexed lobes to expose an elongated yellow pseudocolumella........ ***Physarella oblonga*** (page 124)
9. Sporotheca globose, lenticular or lobed........ 10
10. Sporotheca lobed or somewhat convoluted........ ***Physarum polycephalum*** (page 132)
10. Sporotheca globose or lenticular........ 11
11. Sporotheca lenticular, typically somewhat nodding, stalk noncalcareous........ ***Physarum album*** (page 126)
11. Sporotheca globose, sporangium erect, stalk calcareous........ ***Physarum globuliferum*** (page 130)

Order Stemonitidales

Members of the order Stemonitales, like the Physarales, typically have dark-colored (purple-brown to black) spores, but in a very few examples the spores are ferruginous or dull pink. The predominant type of fruiting body produced by the members of this order is a stalked sporangium; other types such as sessile sporangia, aethalia, pseudoaethalia, and plasmodiocarps are only occasionally encountered. A capillitium is always present, and many species have a prominent columella. The most common expression is for the capillitium to consist of a network of smooth dark threads. The latter may be attached to the top of the stalk, to the columella, to the peridium, or (in one genus) to a disk at the apex of the columella.

The most frequently encountered representatives of this order are species of *Comatricha* and *Stemonitis*, as the concepts are used in traditional taxonomic treatments of the myxomycetes such as the monograph by Martin and Alexopoulos (1969). At the time it appeared, this monograph was the single most important work on the myxomycetes

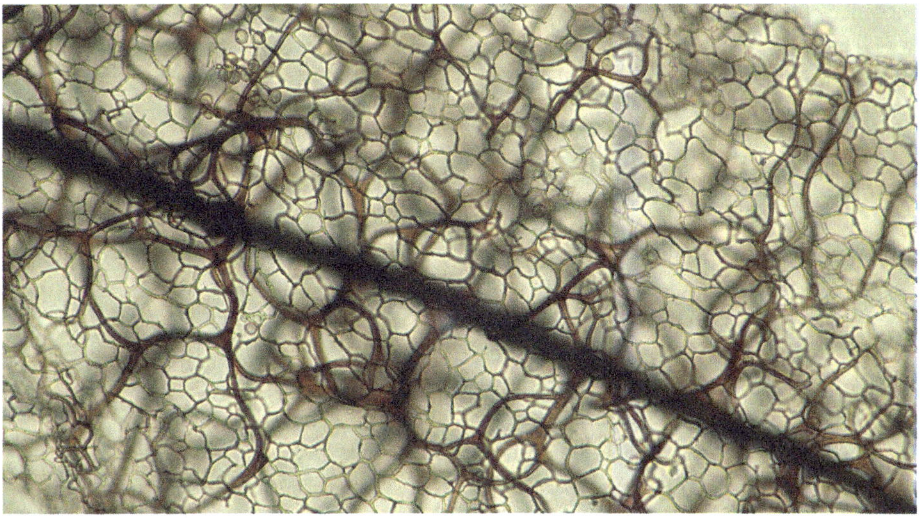

Figure 6-60. Surface net of *Stemonitis smithii* (400x) (©Alain Michaud)

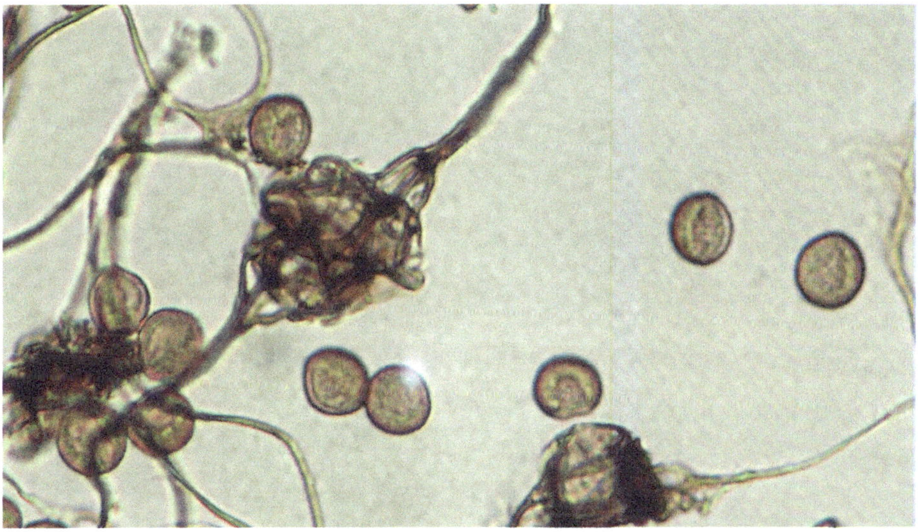

Figure 6-61. Multicellular vesicles of *Brefeldia maxima* (630X) (©Alain Michaud)

ever published. However, detailed studies since then have resulted in many of the species these authors assigned to *Comatricha* and *Stemonitis* (as well as species in other genera) being moved to other "new" segregate genera. For example, what they considered as *Comatricha typhoides* is recognized as *Stemonitopsis typhina* in more recent treatments of the myxomycetes. It is beyond the scope of this book to provide an overview of most of these segregate genera and renamed species. However, many of these are referred to in the comments provided for each of the 58 primary species, and additional information can be obtained from some of the more technical publications on myxomycetes.

Both *Comatricha* and *Stemonitis* (when considered using the concepts of Martin and Alexopoulos) contain a number of common and widespread species, some of which are fairly distinctive. The single most important distinguishing feature of *Stemonitis* is the presence of a surface net, a structure lacking in *Comatricha*. The surface net (Figure 6-60), which develops just beneath the peridium and is connected to the tips of branches of the capillitium, forms a well-defined netlike covering over the surface of the spore mass. However, determining whether or not a surface net is present is not always easy.

Key to common species of the order Stemonitales

1. Surface net present, sporotheca cylindrical with no evidence of a peridium................................2
1. Surface net absent, sporotheca more or less globose or cylindrical with at least some fragments of a peridium present 3
2. Spores bright reddish brown...***Stemonitis axifera*** (page 136)
2. Spores fuscous or dark purple-brown ....***Stemonitis fusca*** (page 137)
3. Stalk long, usually more than 5 times the height of the sporotheca ....***Comatricha nigra*** (page 84)
3. Stalk shorter, much less than 5 times the height of the sporotheca 4
4. Sporangium with a prominent apical disk ....***Enerthenema papillatum*** (page 104)
4. No evidence of an apical disk..***Stemonitopsis typhina*** (page 139)

Figure 6-62. *Stemonitis axifera* (©Kim Fleming)

Figure 6-63. *Stemonitis fusca* (©Todd Elliott)

Figure 6-64. *Comatricha nigra* (©Marius Muessler)

Figure 6-65. *Enerthenema papillatum* (©Kim Fleming)

Figure 6-66. *Stemonitopsis typhina* (©Juan Francisco Moreno)

CHAPTER 7

Descriptions of Species

The preceding keys represent only the first step in identifying a particular specimen. Once an identification as to species is reached, it should be checked against the images (both pen and ink drawings and color images) and description provided for each species included in the keys. An identification can sometimes be made without using the keys by simply thumbing through the illustrations until a match is found, but this is a much less efficient approach and works only for distinctive species. Even if the specimen being identified appears to match the illustration, it should be carefully checked against the information given in the description. Although a few species are so distinctive they are unlikely to be mistaken for anything else, other more closely related species confound even an expert.

If a specimen does not fully agree with the description or the illustration, either a mistake has been made in using the keys (for example, the incorrect lead in a particular couplet may have been selected) or the specimen in question represents a species not included in the keys. Since the keys, illustrations, and descriptions provided herein are limited to 58 of the more common and widespread species, the latter situation should always be considered a distinct possibility. Nevertheless, many species not listed in the keys are discussed in the "Comments" provided with each description because they superficially resemble the species illustrated and described in detail. A color image is provided for 74 additional species in the last section of this book. Consequently, the total number of species that can be identified with the information provided herein actually exceeds 230.

Each of the 58 primary descriptions contains detailed information on structural features likely to be of some value in confirming or rejecting an identification. Two other types of information are included. First, the type of substrate upon which the species in question typically fruits is listed. In addition, anything particularly noteworthy about the distribution, ecology, or taxonomy of the species is mentioned, and other species that are generally similar in appearance are discussed.

All measurements are given in metric units. The following conversion formulas may be helpful to those more familiar with Imperial units:

1 inch = 2.5 centimeters (abbreviated "cm")

1 centimeter (cm) = 10 millimeters ("mm") or approximately 2/5 inch

1 millimeter = 1000 micrometers ("µm") or approximately 1/25 inch

1 micrometer = 1/1000 millimeter or approximately 1/25000 inch

All but two of the pen and ink drawings were done with the aid of a camera lucida and were produced by the late Henry Stempen. Two other drawings (7-28B and 7-44B) were prepared by Dmytro Leontyev. The drawing provided for each species usually includes the habit (that is, the general appearance of the fruiting bodies as they typically occur in nature), one or more spores, and (for those species in which this structure is present) a portion of the capillitium. In some instances, other structures of diagnostic importance are also depicted. Except for the two drawings mentioned above, the scale bar located in the region of the habit sketch is equal to 1 mm, while the scale bar located in the region of the spore(s) and capillitium is equal to 10 µm. For Figure 7-28B the scale bar is equal to 0.1 mm, while for Figure 7-44B, the bar is equal to 0.5 mm. In a few cases, the scale of a drawing is indicated by a number that relates the size of the object as shown in the drawing to its actual size. For example, 175× simply indicates that the object appears 175 times larger than its actual size.

Figure 7-1A. *Arcyria cinerea* (©Henri Koskinen)

Arcyria cinerea (Bull.) Pers.

DESCRIPTION: Sporangia stalked, scattered, gregarious or sometimes united by their fused stalks and occurring in digitate clusters of 2 to 20 or more sporangia, 0.5 to 4.0 mm tall. **Sporotheca** ovate to cylindrical, usually tapering toward the apex, 0.1 to 0.8 mm in diameter. **Stalk** slender, pallid to black, up to 2 mm long. **Hypothallus** discoid or common to a group of sporangia, often inconspicuous. **Peridium** pale gray to light brown, ochraceous or less commonly yellowish brown, persisting in mature sporangia only at the base of the spore mass, where it forms a rather small, shallow calyculus with a smooth, delicately stippled or finely reticulate inner surface. **Capillitium** firmly attached to the calyculus, colorless or nearly so by transmitted light, consisting of a network of threads 2 to 6 µm in diameter and densely covered with small blunt spines (occasionally also with cogs, bands, or reticulations). **Spores** pale gray or light yellow in mass, colorless by transmitted light, with a few scattered warts, 6 to 7 µm in diameter. **Plasmodium** white or less commonly gray or yellowish.

SUBSTRATE: Decaying wood, bark, leaf litter, and dung.

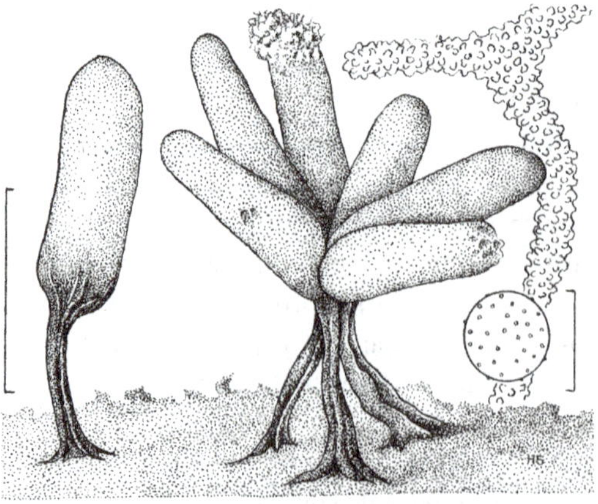

Figure 7-1B. *Arcyria cinerea*. A single sporangium (left side of drawing). A cluster of sporangia with fused stalks (middle). A spore and a portion of the capillitium (right side of drawing).

COMMENTS: This species is relatively common in the field and exceedingly so in moist chamber cultures, especially cultures prepared with samples of leaf litter from broadleaf trees. Two similar species are *Arcyria globosa* Schwein. [page 147, Figure 8-3] and *A. pomiformis* (Leers) Rostaf. [page 148, Figure 8-6]. The former, which has globose sporangia and a deep bowl-shaped calyculus, occurs almost exclusively on old chestnut (*Castanea*) burs. *Arcyria pomiformis*, which appears to be very closely related to *A. cinerea*, can be distinguished on the basis of its generally smaller size, more or less ovate sporangia, yellow spores, and more strongly marked calyculus.

Figure 7-2A. *Arcyria denudata* (©Alain Michaud)

Arcyria denudata (L.) Wettst.

DESCRIPTION: **Sporangia** stalked, usually gregarious to crowded but occasionally scattered, 2 to 6 mm tall. **Sporotheca** ovate to cylindrical, 0.4 to 1.2 mm in diameter. **Stalk** slender, striate, the same color as the sporotheca or darker, 0.5 to 1.5 mm long. **Hypothallus** reddish brown with a silvery sheen but often inconspicuous. **Peridium** pinkish red to brick red but weathering to brown, persisting in mature sporangia only as a shallow calyculus. **Capillitium** consisting of a network of threads 3 to 4 μm in diameter and marked with cogs and half-rings, firmly attached to the whole inner surface of the calyculus. **Spores** red or reddish brown in mass, colorless by transmitted light, with a few scattered warts, 6 to 8 μm in diameter. **Plasmodium** white.

SUBSTRATE: Decaying wood and (less commonly) bark.

COMMENTS: This is one of the more commonly encountered myxomycetes. *Arcyria incarnata* (Pers. ex J.F. Gmel.) Per. [page 148, Figure 8-4], a very similar species, can be distinguished by its capillitium, which is loosely attached only to the center of the calyculus and from which it is readily detached. *Arcyria ferruginea* Saut., another species with a loosely attached capillitium, is much less common than *A. incarnata* and has larger spores (9 to 12 μm in diameter) and capillitium (5 to 8 μm in diameter). (The spores and capillitium of *A. incarnata* are the same size as those of *A. denudata*.) *Arcyria oerstedii* Rostaf. has a conspicuously spiny capillitium, with individual spines sometimes reaching 3 μm in length. *Arcyria stipata* (Schwein.) Lister [page 149, Figure 8-7] is characterized by crowded, more or less superimposed sporangia and faint to strongly marked spiral bands on the threads of the capillitium. Moreover, *A. stipata* tends to be uncommon and typically fruits late in the season.

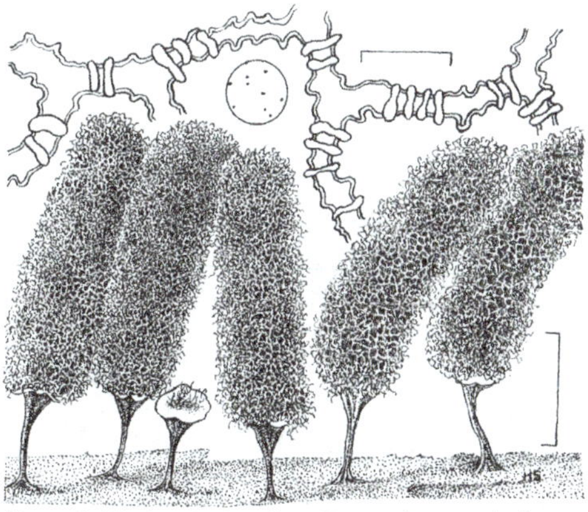

Figure 7-2B. *Arcyria denudata*. Group of sporangia (lower portion of drawing). A single spore and a portion of the capillitium (upper portion of drawing).

Figure 7-3A. *Arcyria versicolor* (©Alain Michaud)

Arcyria versicolor W. Phillips.

DESCRIPTION: **Sporangia** stalked to nearly sessile, gregarious to crowded, 2.0 to 3.5 mm tall at first but increasing to 5 to 6 mm tall after the capillitium expands. **Sporotheca** deeply clavate to pyriform, 1 to 2 mm in diameter. **Stalk** short, weak, filled with spore-like cells, 0.1 to 0.5 mm long. **Hypothallus** prominent, venulose. **Peridium** shining, bright yellow to olivaceous or reddish brown, the basal portion persisting as a deep calyculus with an irregular margin. **Capillitium** strongly elastic, the threads 5 to 6 µm in diameter, spinulose or spiny-reticulate. **Spores** yellow to olivaceous brown or dull reddish brown in mass, pale ochraceous by transmitted light, minutely spiny, 9 to 11 µm in diameter. **Plasmodium** at first white and then becoming orange, coral or yellow.

SUBSTRATE: Decaying coniferous wood, sometimes occurring in snowbank habitats.

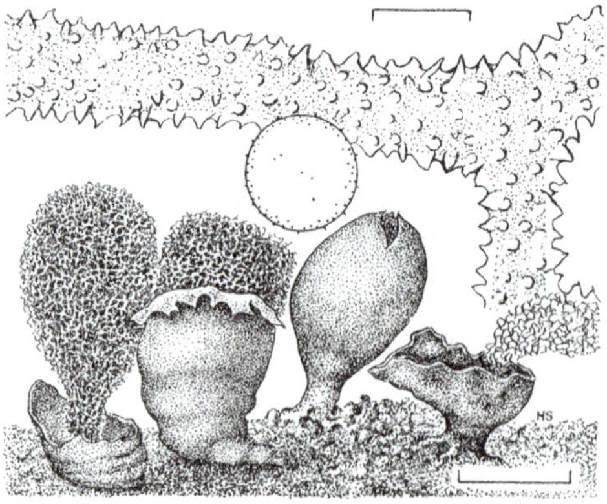

Figure 7-3B. *Arcyria versicolor*. Several sporangia showing various stages of dehiscence (lower portion of drawing). A single spore and a portion of the capillitium (upper portion of drawing).

COMMENTS: *Arcyria versicolor* is found in alpine regions of the temperate zone, where it is often rather common. This species is unlikely to be confused with any other myxomycete. *Arcyria obvelata* (Oeder) Onsberg [page 148, Figure 8-5], a rather common and widely distributed non-alpine species, has generally smaller, bright to pale yellow sporangia and smaller (7 to 8 µm in diameter) spores.

Figure 7-4A. *Badhamia utricularis* (©Henri Koskinen)

Badhamia utricularis (Bull.) Berk.

DESCRIPTION: **Sporangia** stalked (or rarely sessile), gregarious to clustered, 1.0 to 3.0 mm tall. **Sporotheca** globose, ovate or obpyriform, 0.5 to 1.0 mm in diameter. **Stalk** beige, yellow or reddish brown, weak, branched, often prostrate, 2.0 to 4.0 mm long when the sporotheca is pendent. **Hypothallus** thin, dull red. **Peridium** membranous, smooth, rugulose or netted, blue- gray, iridescent violet or cinereous, iridescent, but hyaline or white when the spores have been dispersed. **Capillitium** badhamioid, delicate, consisting of a network of calcareous tubules of a more or less uniform diameter, white. **Spores** dull blackish brown in mass, bright violet brown by transmitted light, loosely aggregated into clusters that readily break apart, warted, 10 to 14 μm in diameter. **Plasmodium** bright yellow.

SUBSTRATE: Dead bark of decaying logs; often associated with old fruiting bodies of wood-decaying fungi.

COMMENTS: Fruitings of several other species of *Badhamia* are likely to be encountered on the bark of decaying logs. Among these are *B. affinis* Rostaf. [page 147, Figure 8-8], *B. capsulifera* (Bull.) Berk., *B. foliicola* Lister [page 149, Figure 8-9], *B. melanospora* Speg. [page 150, Figure 8-10], *B. macrocarpa* (Ces.) Rostaf., *Badhamia nitens* Berk. [page 150, Figure 8-11] and B. *papaveracea* Berk. & Ravenel. *Badhamia capsulifera* and *B. papaveracea* are distinguished from the other species by their spores, which occur in firm clusters of 6 to 20. The sporangia of *B. capsulifera* are sessile or have pale, weak, membranous stalks, while those of *B. papaveracea* typically have relatively well developed, dark stalks. The sporangia of *B. melanospora* are borne on long, weak, pale stalks, while those of *B. affinis*, *B. foliicola*, *B. macrocarpa*, and *B. panicea* [page 150, Figure 8-12] are sessile or very short-stalked. *Badhamia melanospora* also has large (12 to 16 mm in diameter), usually somewhat angular spores. *Badhamia foliicola* has small (usually no more than 0.5 to 0.6 mm in diameter), iridescent gray sporangia that are often somewhat plasmodiocarpous and minutely warted spores that are 11 to 12 μm in diameter. The sporangia of *B. affinis* are pale gray to almost white and more or less flattened or depressed. The spores in this species are densely spinulose and mostly 13 to 15 μm in diameter. *Badhamia macrocarpa* has white, typically globose to subglobose sporangia and irregularly verruculose spores that are 11 to 15 μm in diameter. *Badhamia nitens* differs from all the species listed above by having a yellow peridium.

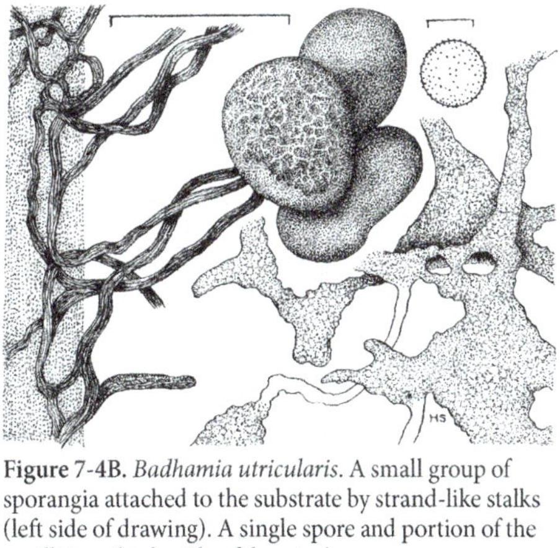

Figure 7-4B. *Badhamia utricularis*. A small group of sporangia attached to the substrate by strand-like stalks (left side of drawing). A single spore and portion of the capillitium (right side of drawing).

Figure 7-5A. *Barbeyella minutissima* (©Alison Pollack)

Barbeyella minutissima Meyl.

DESCRIPTION: **Sporangia** stalked, usually scattered, 0.4 to 0.9 mm tall. **Sporotheca** globose, 0.1 to 0.2 mm in diameter. **Stalk** blackish brown, filled with dark granules, 0.5 to 0.7 mm long. **Hypothallus** inconspicuous. **Peridium** violaceous to brownish black, persistent, somewhat shining, impregnated with dark granules (especially above), dehiscing by five to nine lobes. **Columella** extending into middle of the spore mass and there giving rise to seven to ten simple or forked capillitial threads. **Spores** purple-brown in mass, smoky brown by transmitted light, warted to nearly smooth, 7 to 9 µm in diameter. **Plasmodium** at first colorless, then becoming white, and finally dark brown.

SUBSTRATE: Decaying spruce (*Picea*) logs; almost always associated with leafy liverworts.

COMMENTS: Because of its very small size, *Barbeyella minutissima* is difficult to spot in the field. The best way to "collect" this species is to examine, under magnification in the home or laboratory, pieces of decaying spruce wood bearing a dense growth of leafy liverworts (Smith and Stephenson 2007).

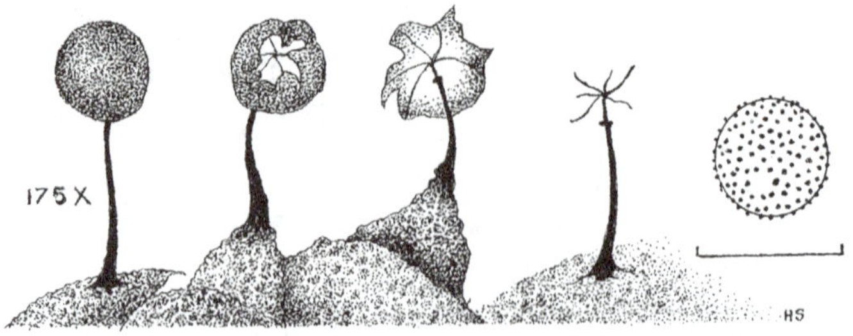

Figure 7-5B. *Barbeyella minutissima*. Intact and dehisced sporangia on the leaves of a leafy liverwort (left and center of drawing). A single spore (right side of drawing).

Figure 7-6A. *Ceratiomyxa fruticulosa* (poroid form) (©Kim Fleming)

Ceratiomyxa fruticulosa (O.F. Müll.) T. Macbr.

DESCRIPTION: Fruiting bodies unlike those of other myxomycetes, typically consisting of a series of erect, simple or branched columns but sometimes taking the form of a poroid or effused structure, usually white but sometimes pink or pale yellow to yellowish green, often 10 cm or more in extent, with the individual columns up to 10 mm high. **Spores** attached individually by thread-like stalks, white in mass, hyaline by transmitted light, smooth, typically somewhat variable in size and shape, mostly ovoid or elliptical but sometimes globose, 6 to 7 × 10 to 13 µm in diameter. **Plasmodium** watery and translucent, colorless or occasionally very pale yellow or more rarely somewhat pink.

SUBSTRATE: Decaying wood; occasionally occurring on dead leaves and other plant debris on the forest floor.

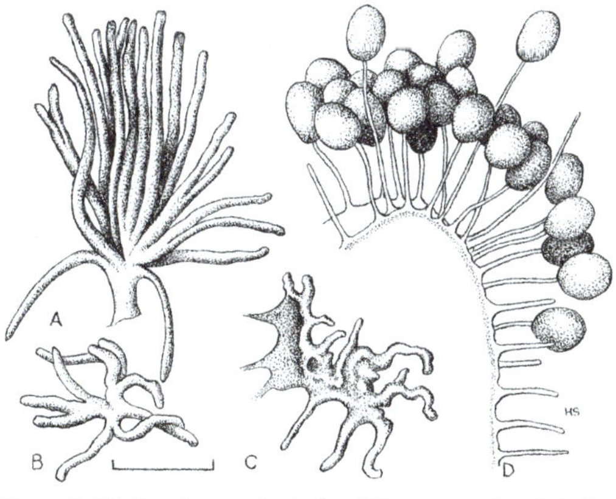

Figure 7-6B. *Ceratiomyxa fruticulosa* (filamentous form). A-C, Three small fruiting bodies showing the variation that exists for this species. D. A semidiagrammatic sketch of a small portion of the surface of one of the erect columns showing spores attached individually by threadlike stalks.

Figure 7-6C. *Ceratiomyxa fruticulosa* (filamentous form) (©Kim Fleming)

COMMENTS: This is one of the most distinctive, abundant, and widespread myxomycetes. *Ceratiomyxa sphaerosperma* Boedijn [page 152, Figure 8-16], a tropical species, can be distinguished by its much smaller size (only 1 to 2 mm tall) and generally smaller (7 to 9 µm in diameter), slightly roughened spores. *Ceratiomyxa morchella* A.L. Welden [page 151, Figure 8-15], another tropical species, has a fruiting body that resembles a tiny morel. Fruitings of *Ceratiomyxa fruticulosa* that take the form of a poroid or effused structure have been recognized as *C. fruticulosa* var. *porioides* (Alb. & Schwein.) G. Lister. In some regions, the latter is relatively common, but in others it is rarely encountered.

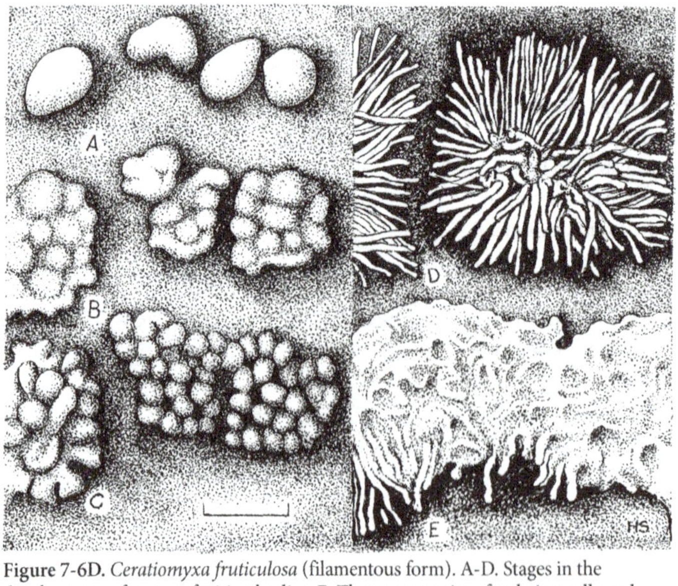

Figure 7-6D. *Ceratiomyxa fruticulosa* (filamentous form). A-D. Stages in the development of mature fruiting bodies. E. The same species after being collected and allowed to air-dry overnight.

Figure 7-7A. *Clastoderma debaryanum* (©Steve Young)

Clastoderma debaryanum A. Blytt

DESCRIPTION: **Sporangia** stalked, usually scattered, 1.0 to 1.3 mm tall. **Sporotheca** slightly nodding, globose, 0.1 to 0.2 mm in diameter. **Stalk** slender, filled with granular material below but translucent and brown above, with the two portions usually separated by a prominent oval swelling, 0.9 to 1.0 mm long. **Hypothallus** inconspicuous. **Peridium** persisting in mature sporangia only as a basal collar, spore mass brown before dispersal. **Columella** short or lacking. **Capillitium** arising from the apex of columella or the base of the sporotheca and consisting of a network of pale brown thread-like elements, with some of the ends expanded to form small platelets. **Spores** bright rosaceous brown in mass, pallid by transmitted light, covered with small warts, 8 to 10 μm in diameter. **Plasmodium** at first watery white and then becoming darker with age.

COMMENTS: The characteristic structure of the stalk makes *Clastoderma debaryanum* an easy species to identify. Three other species of *Clastoderma* have been described, but these are known from only a few collections and are not likely to be encountered by the average collector. The only possible exception is *C. pachypus* Nann.-Bremek., which occasionally occurs in moist chamber cultures on bark. This species is much smaller (no more than 0.5 mm tall) than *C. debaryanum* and the stalk has a very broad base. It was once thought that the peridium of *C. debaryanum* persisted in a mature sporangium only as fragments attached to the tips of capillitial threads. However, Frederick et al. (1986) reported that what appear to be fragments of the peridium are actually flattened, membranous structures continuous with the capillitium and do not represent remnants of the peridium, which disappears completely from mature sporangia.

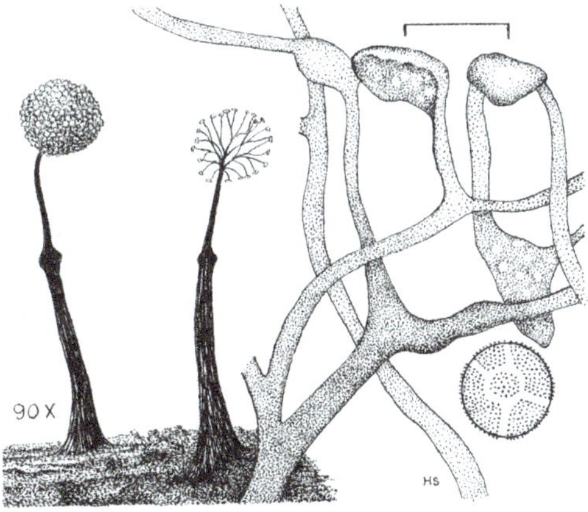

Figure 7-7B. *Clastoderma debaryanum*. Two sporangia, one with the spore mass still intact and the other with the spores mostly lost (left side of drawing). A single spore and portions of the capillitium with attached scales derived from the peridium (right side of drawing).

Figure 7-8A. *Comatricha nigra* (©Marius Muessler)

Comatricha nigra (Pers. ex J.F. Gmel.) J. Schröt.

DESCRIPTION: **Sporangia** stalked, scattered or gregarious, 1 to 8 mm tall. **Sporotheca** erect, globose or ovate (rarely short-cylindrical), dark brown or black, up to 0.6 mm in diameter. **Stalk** black, hair-like, relatively long, usually two to six times the diameter of the sporotheca. **Hypothallus** red, sparse. **Columella** reaching to the middle or the upper part of the sporotheca, there merging into the capillitium. **Peridium** not persisting in mature sporangia, spore mass dark brown or black before dispersal. **Capillitium** an intricate network of slender, flexuous, branching and anastomosing purplish brown threads. **Spores** dark brown to black in mass, dark violaceous by transmitted light, faintly warted to nearly smooth, 9 to 10 µm in diameter. **Plasmodium** transparent and then becoming white.

SUBSTRATE: Decaying wood; also appearing on tree bark in moist chamber cultures.

COMMENTS: In the field, the combination of a small globose sporotheca on a long, slender stalk makes this an easy species to recognize. One morphologically similar species *is Comatricha elegans* (Racib.) G. Lister, but the columella in this species is absent or extremely short and gives rise to the branches of the capillitium at or near the base of the sporangium. *Comatricha ellae* Härk. is superficially similar to *C. elegans*, but the former has a more or less complete surface net at the surface of the sporotheca. Several other similar species were once placed in the genus *Comatricha* but have since been reassigned to other genera. All of these have a globose sporotheca and a relatively long stalk, and they commonly appear on tree bark in moist chamber cultures. Some of these species are not easy to identify.

Collaria lurida (Lister) Nann.-Bremek. is typically smaller than *Comatricha nigra*, has a relatively longer stalk, and occurs on dead leaves and not bark. *Paradiacheopsis acanthodes* (Alexop.) Nann.-Bremek., *Paradiacheopsis solitaria* (Nann.-Bremek.) Nann.-Bremek. [page 164, Figure 8-52], and *P. fimbriata* (G. Lister & Cran) Hertel ex Nann.-Bremek. do occur on bark but differ from *C. nigra* in a number of respects. *Paradiacheopsis acanthodes* has large (12 to 13 µm in diameter), conspicuously spiny spores, while *P. fimbriata* has an unusually scanty capillitium in which the individual branches have expanded tips. In *P. solitaria* the spores are large (typically 14 to 16 µm in diameter) but verruculose and not spiny. Moreover, the capillitium is well developed in *P. solitaria*. A key to some of the other species of *Comatricha* and other morphologically similar forms that commonly appear in moist chamber cultures was provided by Mitchell (1980).

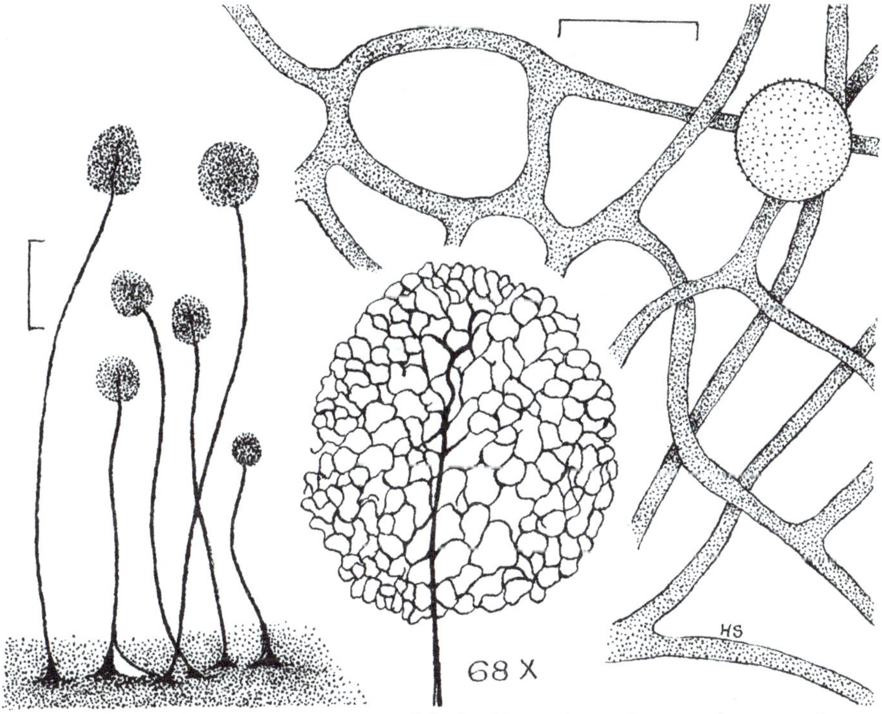

Figure 7-8B. *Comatricha nigra*. Group of sporangia (left side of drawing). A single spore and a portion of the capillitium (upper right). A semidiagrammatic sketch of a single sporangium showing the relationship of the capillitium to the columella (lower center).

Figure 7-9A. *Craterium minutum* (©Alain Michaud)

Craterium minutum (Leers) Fr.

DESCRIPTION: **Sporangia** stalked (or rarely sessile), gregarious, 0.3 to 1.5 mm tall. **Sporotheca** erect, goblet-shaped to subcylindrical, 0.2 to 0.8 mm in diameter. **Stalk** usually slightly paler than the base of the cup, orange-red, translucent, furrowed, half the total height or less. **Hypothallus** discoid, translucent orange-brown. **Peridium** ochraceous brown or olivaceous to deep chocolate, umber, or bright brownish red, usually darker below, thick, consisting of two layers, the outer layer cartilaginous, the inner layer limy and white, dehiscence by a distinct preformed operculum that is sharply separated from the rest of the peridium. **Capillitium** physaroid, the nodes large, irregular, white or ochraceous, tending to become aggregated at the center. **Spores** black in mass, violaceous brown by transmitted light, minutely warted, 8 to 10 µm in diameter. **Plasmodium** white to orange-yellow or orange-red

SUBSTRATE: Dead leaves and twigs.

COMMENTS: Other commonly encountered species of *Craterium* lack a distinct preformed operculum. Sporangia of *C. leucocephalum* (Pers. ex J.F. Gmel.) Ditmar [page 154, Figure 8-22] are white or almost so above, those of *C. aureum* (Schumach.) Rostaf. [page 153, Figure 8-21] are bright yellow or greenish yellow, and those of C. *paraguayense* (Speg.) G. Lister are pinkish to deep purple. *Craterium obovatum* Peck [page 154, Figure 8-23], which is assigned to the genus *Badhamia* as *B. obovata* (Peck) S.J. Sm. in some texts, has grayish brown to purplish brown sporangia and a more or less badhamioid capillitium. It can be distinguished from *C. paraguayense*, which also has purplish brown sporangia, by the color of the capillitium. The former species has a white capillitium, while the latter species has a violet capillitium. The basal portion of the peridium of *Physarum pusillum* (Berk. & M.A. Curtis) G. Lister [page 165, Figure 8-63] tends to persist as a cup-like structure, causing the sporangia to generally resemble the dehisced sporangia of *C. leucocephalum*. The two species also characteristically occur on the same types of substrates (usually dead leaves and other plant debris on the forest floor). *Physarum pusillum* has longer stalks (usually more than half the total height as opposed to one-half or less in *C. leucocephalum*) and larger spores (10 to 12 µm in diameter for *P. pusillum* versus 8 to 9 µm in diameter for *C. leucocephalum*).

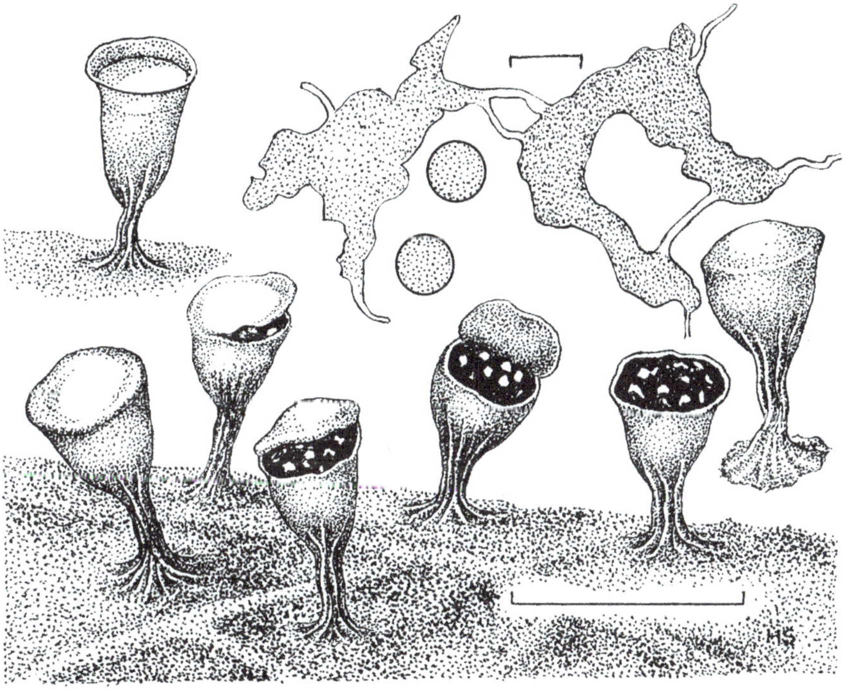

Figure 7-9B. *Craterium minutum*. Several sporangia on a dead leaf (lower portion of drawing). Two spores and a portion of the capillitium (upper right). A single sporangium showing a distinct hypothallus (far right).

Figure 7-10A. *Cribraria argillacea* (©Alain Michaud)

Cribraria argillacea (Pers. ex J.F. Gmel.) Pers.

DESCRIPTION: **Sporangia** stalked (rarely sessile to short-stalked), usually closely gregarious to densely crowded (sometimes to the point of forming a pseudoaethalium), 0.75 to 1.50 mm tall. **Sporotheca** erect, globose to obovate, 0.5 to 1.0 mm in diameter. **Stalk** furrowed, dark brown to black, 0.1 to 0.5 long. **Hypothallus** well developed, brown. **Peridium** dull ochraceous to clay-colored or olivaceous, persisting below as a deep, often poorly defined calyculus and above as a weak, easily detached peridial net without thickened nodes. **Spores** clay-colored in mass, pallid by transmitted light, nearly smooth, 6 to 8 µm in diameter. **Plasmodium** lead-colored.

SUBSTRATE: Decaying wood, especially that of coniferous trees.

COMMENTS: The most distinctive characteristics of this species are the relatively large, sessile to short-stalked, clay-colored sporangia and the weak, poorly developed peridial net. *Cribraria ferruginea* Meyl. and *Cribraria purpurea* Schrad. are somewhat similar in size and also typically occur on decaying coniferous wood, but these two species are easily distinguished from *C. argillacea* on the basis of color. *Cribraria purpurea* has reddish purple sporangia and a peridial net with conspicuously expanded nodes, while *C. ferruginea* is brick red to deep purple-red and has a peridial net with scarcely differentiated nodes. *Cribraria rufa* (Roth) Rostaf. [page 155, Figure 8-26], another species typically associated with decaying coniferous wood, has reddish orange sporangia and a rather large meshed peridial net. Fruitings of *C. argillacea* in which the sporangia are densely crowded often resemble the aethaliate to pseudoaethaliate fruiting bodies produced by *Lindbladia tubulina* Fr., but the former can be distinguished by its more delicate peridium and at least some evidence of a definite peridial net. Moreover, *L. tubulina* has a tough, dark, broadly effused and continuous hypothallus that is not present in *C. argillacea*.

Figure 7-10B. *Cribraria argillacea*. Several sporangia showing the variation that exists in this species (lower portion of drawing). A single spore and a portion of the peridial net showing dictydine granules (upper portion of drawing).

Figure 7-11A. *Cribraria cancellata* (©Yuri Novozhilov)

Cribraria cancellata (Batsch.) Nann.-Bremek.

DESCRIPTION: **Sporangia** stalked, gregarious, often occurring in extensive fruitings, 1 to 5 mm tall. **Sporotheca** usually nodding, globose, 0.4 to 0.7 mm in diameter. **Stalk** long, slender, striate, often twisted near the apex, dark brown to black below, one to eight times as long as the diameter of the sporotheca. **Hypothallus** inconspicuous. **Peridium** deep reddish brown or brownish purple, varying to bright purple or rufous membranous, delicate, persisting in mature sporangia as a system of rigid, longitudinal ribs connected by delicate transverse filaments. **Spores** brick red or reddish purple-brown in mass, nearly colorless to pale red by transmitted light, nearly smooth to minutely verrucose, 4 to 7 µm in diameter. **Plasmodium** purple-black.

SUBSTRATE: Decaying wood, particularly that of coniferous trees.

COMMENTS: The distinctive cage-like structure of the sporangium makes this very common and widespread species one of the easiest myxomycetes to recognize. *Cribraria mirabilis* (Rostaf.) Massee, a snowbank species, can be distinguished by its darker color and a peridial net that is strongly ribbed below and netted above. In earlier treatments of the myxomycetes, *Cribraria cancellata* is listed as *Dictydium cancellatum* (Batsch) T. Macbr. and *C. mirabilis* is listed as *D. mirabilis* (Rostaf.) Meyl.

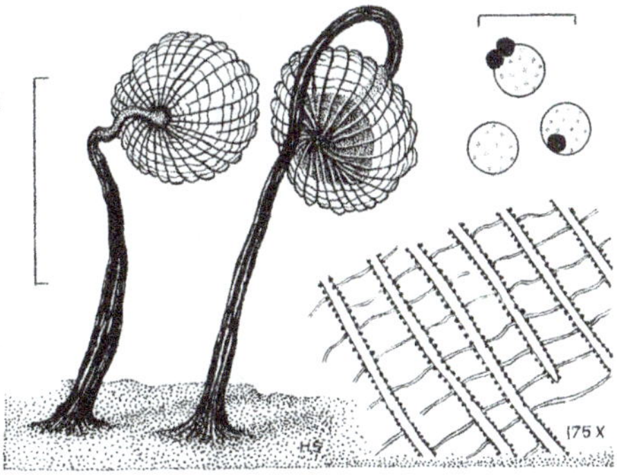

Figure 7-11B. *Cribraria cancellata.* One sporangia has lost all its pores and the other still contains a ball of spores (left side of drawing). Detailed view of the surface net (lower right). Spores and dictydine granules (upper right).

Figure 7-12A. *Cribraria intricata* (©Alain Michaud)

Cribraria intricata Schrad.

DESCRIPTION: Sporangia stalked, gregarious, 1.5 to 3.5 mm tall. **Sporotheca** usually nodding but sometimes erect, globose, 0.5 to 0.7 mm in diameter. **Stalk** slender, furrowed, tapering upward, dark brown, 1.5 to 3.0 mm long. **Hypothallus** inconspicuous. **Peridium** commonly ochraceous to dusky brown to blackish brown, persisting as a peridial net both above and below or forming a calyculus

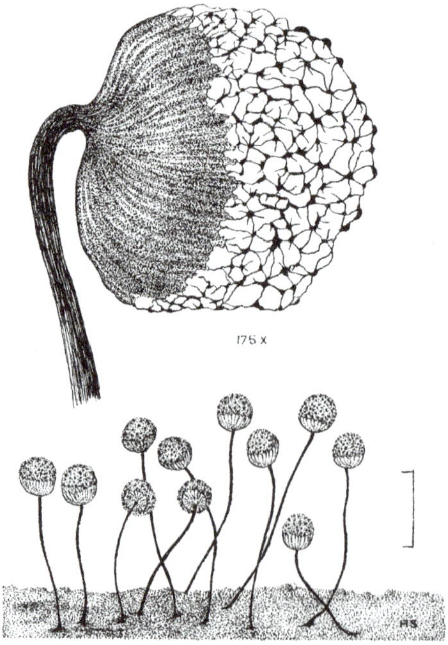

Figure 7-12B. *Cribraria intricata*. Group of sporangia (lower portion of drawing). A single sporangium showing the peridial net and clearly defined calyculus (upper portion of drawing)

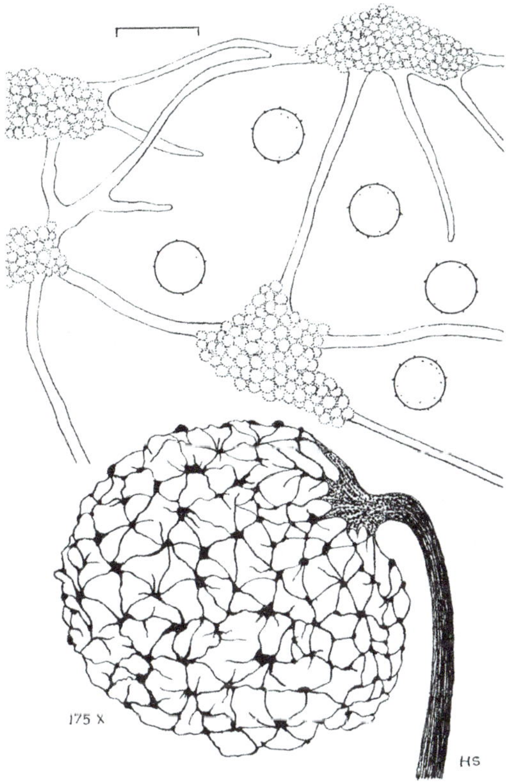

Figure 7-12C. *Cribraria intricata*. A single sporangium after the spores have been lost (lower portion of drawing). Note that the peridial net is well developed but a clearly defined calyulus is absent. A detailed view of a portion of the peridial net showing the attachment of dictyodine granules to the nodes (upper portion of drawing).

below and netted above, meshes of peridial net regular, often somewhat triangular, the nodes dark, prominent, thickened, polygonal or branched, with each node giving rise to five to eight connecting threads and one or more free ends. **Spores** ochraceous in mass, pallid by transmitted light, spinulose, 5 to 6 µm in diameter. **Plasmodium** greenish, lead-colored or brownish black.

SUBSTRATE: Decaying wood.

COMMENTS: This is perhaps the most common large *Cribraria* and often occurs in large fruitings. Two similar species are *C. tenella* Schrad. and *C. splendens* (Schrad.) Pers. *Cribraria tenella* has generally smaller sporangia with more rounded nodes that give rise to four to five connecting threads and few if any free ends. *Cribraria splendens* lacks a calyculus; this structure is replaced by 8 to 15 distinct firm ribs that radiate from the stalk. *Cribraria aurantiaca* Schrad. has a bright yellow spore mass and a small-meshed peridial net with small, dark, rounded nodes. *Cribraria macrocarpa* Schrad. is yellow-brown to light brown or bronze in color and has a calyculus with a perforated upper margin that merges gradually into the well developed peridial net. The sporangia of *C. piriformis* Schrad. are typically dark purple-brown and have rather large, dark dictydine granules in the calyculus and in the nodes of the peridial net. Although *Cribraria* is usually a fairly easy genus to recognize, the same cannot be said for most of the species. Many of the characters used to distinguish one species from another exhibit considerable variation, even within a single fruiting. Identification of species is often possible only when fresh, well-developed material is available for study.

Figure 7-13A. *Cribraria microcarpa* (©Marius Muessler)

Cribraria microcarpa (Schrad.) Pers.

DESCRIPTION: **Sporangia** stalked, scattered to gregarious, 3 to 4 mm or more tall. **Sporotheca** usually nodding, globose, 0.1 to 0.3 mm in diameter. **Stalk** slender, dark purplish brown, usually six or more times as long as the diameter of the sporotheca. **Hypothallus** inconspicuous. **Peridium** reddish ochraceous, persisting as a peridial net both above and below, with a calyculus lacking or represented by a small basal disk, meshes of net mostly rectangular, the nodes rounded, strongly thickened, 10 to 20 µm in diameter. **Spores** ochraceous in mass, pale by transmitted light, minutely spinulose, 5 to 7 µm in diameter. **Plasmodium** at first colorless, then becoming white, and finally dingy brown.

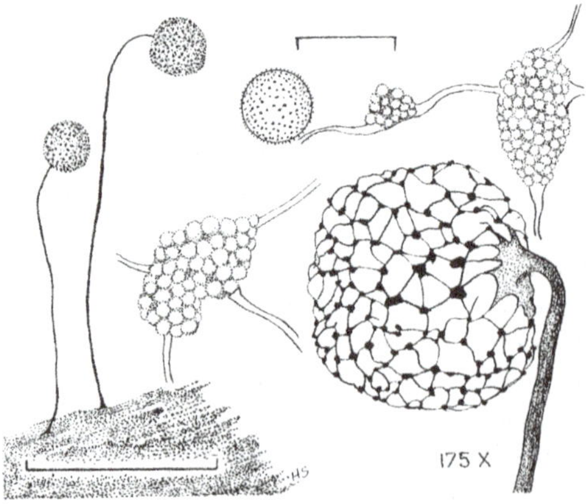

Figure 7-13B. *Cribraria microcarpa*. Two sporangia (left side of drawing). A single sporangium from which the spores have been lost (lower right). One node of the peridial net (middle). A single spore and a portion of the peridial net (upper right).

SUBSTRATE: Decaying wood; also appearing on coniferous leaf litter placed in moist chamber cultures.

COMMENTS: This is the most common species of *Cribraria* with small nodding sporangia and very long delicate stalks. *Cribraria confusa* Nann.-Bremek. & Y. Yamam. [page 154, Figure 8-24] and *C. minutissima* Schwein. [page 155, Figure 8-25] have relatively shorter stalks, nodes that are thickened, and yellow to orange-brown spores. *Cribraria minutissima* has a calyculus, while this structure is lacking in *C. confusa*. The sporangia of *C. violacea* Rex have a deep calyculus and are deep purple in color.

Figure 7-14A. *Diachea leucopodia* (©Alain Michaud)

Diachea leucopodia (Bull.) Rostaf.

DESCRIPTION: **Sporangia** stalked (rarely sessile), gregarious, 0.7 to 2.0 mm tall. **Sporotheca** erect, ellipsoid to cylindrical, 0.3 to 0.6 mm in diameter. **Stalk** stout, brittle, tapering upward calcareous, filled with granular lime, white, one-fourth to one-half the total height of the sporangium. **Hypothallus** white, calcareous, venulose, usually forming a conspicuous network from which the fruiting bodies arise, sometimes sparse. **Peridium**, membraneous iridescent blue, purple, or bronze. *Columella* cylindrical or slightly tapered upward, white, calcareous, one-half or more of the total height of the sporangium. **Capillitium** consisting of branching and anastomosing flexuous threads, brown except at the pale extremities, arising from all parts of the columella. **Spores** nearly black in mass, dull lilaceous by transmitted light, minutely roughened, 8 to 11 µm in diameter. **Plasmodium** white.

SUBSTRATE: Dead leaves, twigs, and other plant debris on the forest floor; also commonly fruiting on living plants.

COMMENTS: *Diachea leucopodia* is a distinctive species and is not likely to be confused with any other myxomycete. *Diachea subsessilis* Peck [page 155, Figure 8-27] is similar in general appearance but has a globose sporotheca and reticulate spores. Two other species of *Diachea* are occasionally encountered. In *D. thomasii* Rex the lime present in the fruiting body is orange in color, while *D. splendens* Peck has globose, iridescent blue sporangia and strongly warted spores. *Elaeomyxa cerifera* (G. Lister) Hagelst. [page 160, Figure 8-40] is morphologically similar to *D. leucopodia*. Both species have an iridescent peridium, but in *E. cerifera* the stalk is dark and not white. Moreover, the stalk of *E. cerifera* has a thick waxy collar at the top, a feature not found in other myxomycetes.

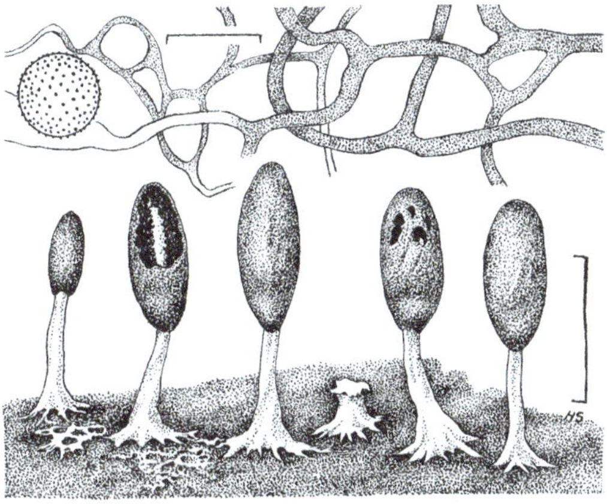

Figure 7-14B. *Diachea leucopodia*. Several sporangia (lower portion of drawing). A single spore and a portion of the capillitium (upper portion of drawing).

Figure 7-15A. *Dictydiaethalium plumbeum* (©Kim Fleming)

Dictydiaethalium plumbeum (Schumach.) Rostaf.

DESCRIPTION: **Pseudoaethalia** (occasionally sporangiate or aethaliate fruiting bodies are encountered) composed of numerous cylindrical sporangia, these 0.3 to 0.5 mm in diameter and 1 to 3 mm tall, sporangia closely compacted into a flat, palisade-like layer, sometimes 10 cm or more in total extent. **Hypothallus** thick (spongy in pseudoplasmodial forms), often extending beyond the margin of the fruiting body, silvery or pallid. **Peridium** membranous, dull yellow to yellowish or dark reddish brown, dark olivaceous, or slate-colored, membranous, not persisting in mature fruiting bodies except at the apices of the individual sporangia (where it is thicker and forms a definite cortex over the upper surface of the fruiting body) and in the junctions between adjacent sporangia (where it persists as thread-like strands), dehiscence occurring as a result of the breaking apart of the individual sporangial "caps," each of which remains connected to the thread-like strands formed in the sporangial junctions. **Hypothallus** thick, often extending beyond the margin of the fruiting body, silvery or pallid. **Spores** olivaceous brown to yellow or clay-colored in mass, nearly colorless to pale yellow by transmitted light, minutely roughened, 8.5 to 11.0 μm in diameter. **Plasmodium** black.

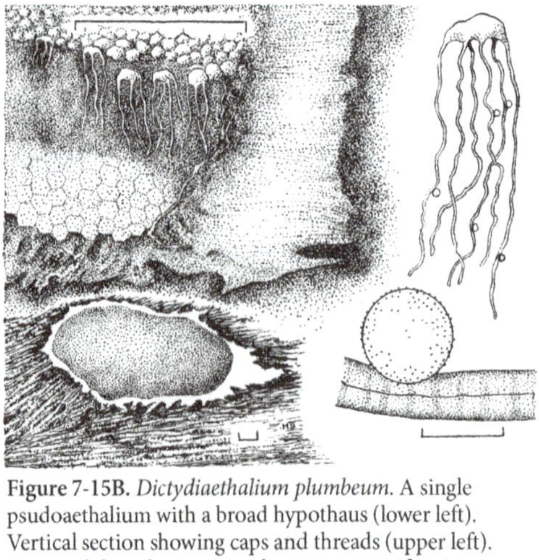

Figure 7-15B. *Dictydiaethalium plumbeum.* A single psudoaethalium with a broad hypothaus (lower left). Vertical section showing caps and threads (upper left). Caps and threads as seen with greater magnification (upper right). A single spore lying next to a portion of a thread (lower right).

SUBSTRATE: Decaying wood.

COMMENTS: Although quite variable in size and color, this species is usually fairly easy to recognize because of the distinctive nature of the peridium.

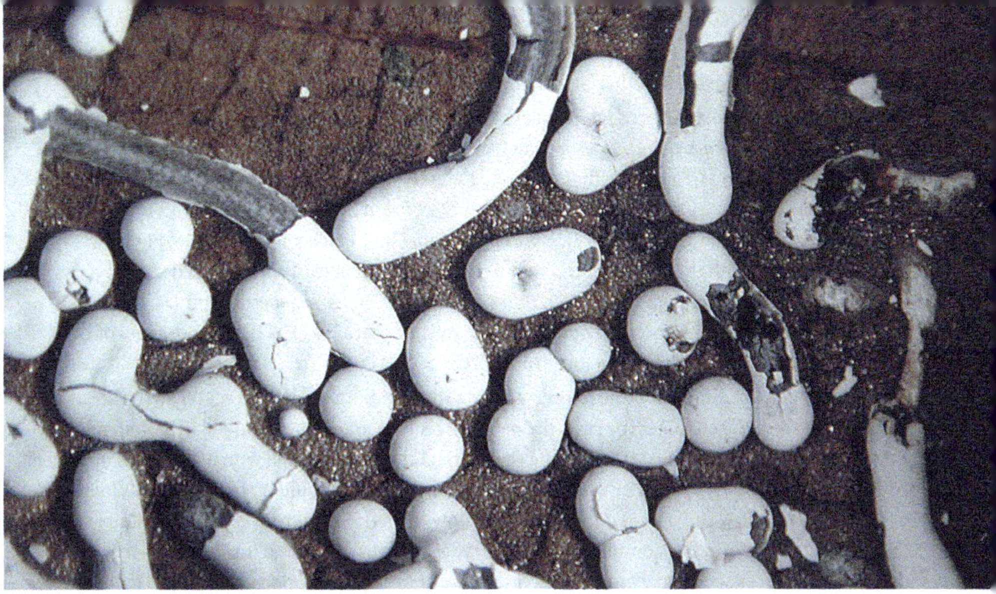

Figure 7-16A. *Diderma effusum* (©Laurie Leonard)

Diderma effusum (Schwein.) Morgan

DESCRIPTION: **Sporangia** sessile or plasmodiocarps, the latter usually broadly effused, gregarious, 0.5 to 1.5 mm broad and up to 6 cm in maximum extent. **Hypothallus** membranous, white, usually not extending beyond the fruiting body and often inconspicuous. **Peridium** white, consisting of two layers, the outer layer calcareous, relatively smooth, the inner layer membranous, colorless. *Columella* flat pulvinate, pinkish brown. **Capillitium** delicate, the threads colorless or pale purple, sparingly branched and anastomosing. **Spores** dark purple in mass, pale violaceous brown by transmitted light, asperulate to verruculose, 7 to 9 µm in diameter. **Plasmodium** white.

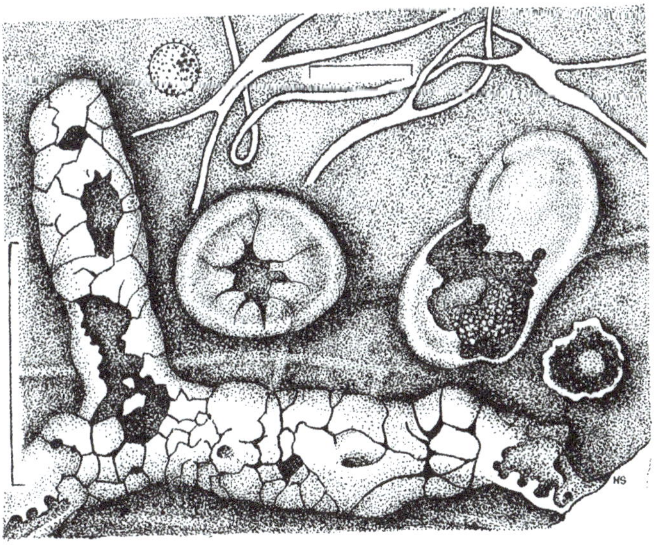

Figure 7-16B. *Diderma effusum*. A typical plasmodiocarp (lower portion of drawing). Several sporangia (middle of drawing). Asingle spore and portions of the capillitium (upper portion of drawing).

SUBSTRATE: Dead leaves and other plant debris on the forest floor.

COMMENTS: Typical fruitings of *Diderma effusum* are easily recognized by their broadly effused plasmodiocarps. *Diderma chondrioderma* (de Bary & Rostaf.) Kuntze is rather similar in appearance but much less common in nature, although it often appears on tree bark in moist chamber cultures. *Diderma chondrioderma* can be distinguished from *D. effusum* by its larger (mostly 12 to 13 µm in diameter) spores. Sessile sporangia of *D. hemisphaericum* (Bull.) Hornem. [page 156, Figure 8-29] might be confused with those of *D. effusum*, but the presence of at least some stalked sporangia in a given fruiting is enough to distinguish this species from *D. chondrioderma* and *D. effusum*, both of which are sessile. As pointed out by Martin and Alexopoulos (1969), the sporangia of *D. hemisphaericum* resemble miniature mushrooms.

In some species of *Didymium* the lime crystals on the peridium are closely aggregated instead of being scattered. When this is the case, the peridium is firm and smooth, and has the general appearance of the shell of an egg. Unless the peridium is closely examined with a hand lens or under a stereomicroscope, any species in which this condition is the case could be confused with members of the genus *Diderma*. *Didymium difforme* (Pers.) Gray [page 156, Figure 8-35] is one example. This species, which is rather common on various types of plant debris and the dung of herbivorous mammals, produces sporangia and simple plasmodiocarps similar in appearance to those sometimes found in *Diderma effusum*. However, in addition to the difference in the peridium of the two species, *Didymium difforme* has appreciably larger spores (11–14 µm versus 7–9 µm in *Diderma effusum*). *Didymium dubium* Rostaf. [page 158, Figure 8-36] is a sessile species of *Didymium* in which the lime crystals are spread over the peridium. The fruiting bodies of *Didymium dubium* are effused plasmodiocarps and are found in both lowland and alpine habitats, although some collections from alpine habitats previously referred to *D. dubium* have now been referred to other species such as *D. nivicola* Meyl. (Ing 1999). *Didymium anellus* Morgan [page 157, Figure 8-33] is another species that forms effused plasmodiocarps in which the lime crystals are spread over the peridium. The plasmodiocarps of *D. anellus* tend to be small and either somewhat ring-shaped or have a noticeably depressed center. This species typically occurs on dead leaves.

Figure 7-17A. *Diderma floriforme* (©Alain Michaud)

Diderma floriforme (Bull.) Pers.

DESCRIPTION: Sporangia stalked, usually densely crowded and often in large fruitings, 1 to 2 mm tall. **Sporotheca** globose to pyriform, 0.7 to 1.0 mm in diameter. **Stalk** weak, furrowed, orange-brown to dark, merging into the hypothallus, 0.4 to 1.1 mm long. **Hypothallus** prominent, the same color as the stalk. **Peridium** tan to deep reddish brown or occasionally gray, consisting of two layers, the outer layer cartilaginous, usually closely adherent to the membranous inner layer, dehiscence irregular above, the lower portion dividing into petal-like lobes which often become deeply recurved and revolute, the base tending to persist as a calyculus. **Columella** prominent, clavate, globose above,

Figure 7-17B. *Diderma floriforme*. Three sporangia, one intact and two dehisced (lower portion of the drawing). A single spore and portion of the capillitium (upper portion of the drawing).

the surface rugose, densely calcareous, ochraceous to brown. **Capillitium** abundant, the threads delicate, slender, sinuous, dark brown, often bearing bead-like thickenings. **Spores** black in mass, dark purplish brown by transmitted light, paler on one side, sparsely and irregularly marked with large, blunt warts, 10 to 11 µm in diameter. **Plasmodium** yellow or grayish white.

SUBSTRATE: Decaying wood.

COMMENTS: Several other species of *Diderma*, all of which are sessile to short-stalked, are characterized by a similar type of dehiscence. *Lepidoderma trevelyanii* (Grev.) Poulain & Mar. May. (listed as *Diderma trevelyanii* [Grev.] Fr. in earlier treatments of the myxomycetes) [page 161, Figure 8-43] and *D. asteroides* (Lister & G. Lister) G. Lister have a peridium consisting of three layers. The white calcareous middle layer is usually quite conspicuous. In *L. trevelyanii* this middle layer is composed of crystalline lime, while in *D. asteroides* the lime is noncrystalline. The latter species also tends to have smaller (typically less than 0.8 mm in diameter) sporangia than the former (in which the sporangia are usually more than 0.8 mm in diameter). In both species, the peridium splits into lobes that fold back, exposing the white middle layer in striking contrast to the typically darker substrate. The dehisced sporangia often resemble small white stars. *Diderma radiatum* (L.) Morgan [page 157, Figure 8-31] has the same type of stellate dehiscence, but its peridium lacks a prominent white middle layer. *Diderma radiatum* differs from *D. floriforme* by having more delicately marked (minutely verrucose or spinulose) spores. *Diderma montanum* (Meyl.) Meyl. [page 156, Figure 8-30] is another short-stalked member of the genus in which the sporotheca is subglobose to hemispherical and often somewhat flattened. However, unlike the other species mentioned above, dehiscence is irregular.

As mentioned under *Diderma effusum*, some species of *Didymium* have a peridium that is superficially similar in appearance to that found in members of the genus *Diderma*. One of these is *Didymium vaccinum* (Durieu & Mont.) Buchet page 159, Figure 8-39], a species associated with desert habitats, where it is often one of the more common myxomycetes present. The sporangia of *Didymium vaccinium* can be sessile but more commonly have a short stout stalk. The upper portion of the sporotheca has the shape of a hemisphere, and when this falls away during dehiscence, it leaves behind a cup-like structure with a conspicuous columella arising from the middle.

Figure 7-18A. *Diderma niveum* (©Alain Michaud)

Diderma niveum (Rostaf.) E. Sheld.

DESCRIPTION: **Sporangia** sessile, gregarious to (more commonly) crowded, usually depressed globose to pulvinate but sometimes elongated and plasmodiocarpous, 0.7 to 2.2 mm in diameter. **Hypothallus** white, often abundant, sometimes scanty or lacking. **Peridium** white or pale pinkish beige, double, the outer layer crustaceous, chalky, smooth, fragile, the inner layer delicate, persistent, membranous and often iridescent, yellow or orange below. **Columella** large, globose or hemispherical, ochraceous to deep orange or sometimes white. **Capillitium** abundant, elastic, the threads of two types, some purplish or dusky, coarse, uneven, with pale extremities, others delicate and colorless, often beaded with wart-like thickenings, both types rather sparsely branched and anastomosing. **Spores** black in mass, violet-brown by transmitted light, minutely roughened, 9 to 11 µm in diameter. **Plasmodium** white.

SUBSTRATE: Dead leaves, twigs, dead bark, and other types of plant debris; sometimes fruiting on decaying wood and living plants.

COMMENTS: This snowbank species is exceedingly common in mountainous regions of the temperate zone. *Diderma fallax* (Rostaf.) E. Sheld. [page 156, Figure 8-28] and *D. montanum* (Meyl.) Meyl. are found in similar situations. The latter has sporangia with short, stout stalks and spores that are minutely spinulose, while the former is either sessile or short-stalked and has large (14 to 17 µm in diameter), dark, conspicuously warted spores.

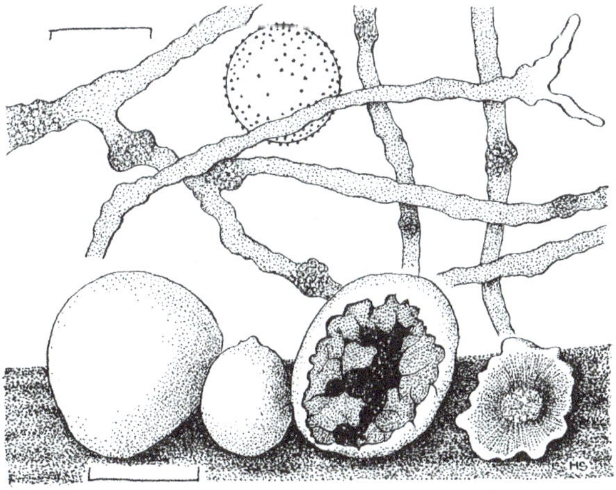

Figure 7-18B. *Diderma niveum.* Intact and dehisced sporangia (lower portion of drawing). Note that a distinct columella is evident in one of the latter. A spore and portions of the capillitium (upper portion of drawing).

Figure 7-19A. *Diderma testaceum* (©Alain Michaud)

Diderma testaceum (Schrad.) Pers.

DESCRIPTION: **Sporangia** sessile (or very rarely short-stalked), gregarious to crowded, hemispherical to flat-pulvinate or depressed globose, erect, pale flesh-colored or pink, 0.7 to 1.0 mm in diameter. **Stalk** (when present) very short, stout, same color as the peridium. **Hypothallus** inconspicuous. **Peridium** pale flesh-colored or pink, double, the outer layer smooth, porcelain-like and composed of a crust of globose lime granules, pinkish within, the inner layer separate but close, thin, membranous, often wrinkled, ashen or pinkish gray. **Columella** prominent, convex, slightly roughened, pink to reddish brown. **Capillitium** abundant, the threads delicate, smooth, sparingly branched, pale or colorless. **Spores** black in mass, light brown by transmitted light, nearly smooth, 8 to 9 µm in diameter. **Plasmodium** yellowish buff.

SUBSTRATE: Dead leaves, twigs, and other plant debris on the forest floor.

COMMENTS: *Diderma testaceum* is one of a number of species of *Diderma* typically associated with forest floor leaf litter. Among the other species found in this microhabitat are *D. cinereum* Morgan, *D. crustaceum* Peck, *D. globosum* Pers., and *D. spumarioides* (Fr. & Palmquist) Fr. [page 147, Figure 8-32]. All four species produce sessile, white sporangia and often are difficult to distinguish from one another. *Diderma cinereum* is probably the least common. It typically fruits rather early in the season and has a thin, rather fragile peridium that appears to consist of a single layer and a colorless, inconspicuous hypothallus. The three other species have a peridium consisting of two distinct layers and a conspicuous (and sometimes rather profuse) white hypothallus. The hypothallus of both *D. crustaceum* and *D. spumarioides* typically is so profuse that the sporangia appear to be more or less embedded in it, which is not the case for *D. globosum*. *Diderma crustaceum* has relatively large (mostly 12 to 14 µm in diameter), dark, spiny spores, while *D. globosum* has dark, strongly but sparsely warted spores that are 10 to 12 µm in diameter. *Diderma spumarioides* has pale, minutely warted spores that are mostly 8 to 11 µm in diameter.

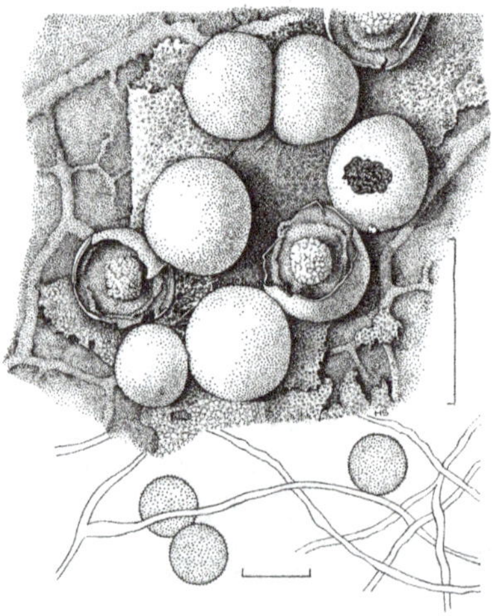

Figure 7-19B. *Diderma testaceum*. Several sporangia on a dead leaf (upper portion of drawing). Note that the columella and the two layers of the peridium are evident in those sporangia in which the peridium has been disrupted. Several spore and portions of the capillitium (lower portion of drawing).

Figure 7-20A. *Diderma iridis* (©Joe Takano)

Didymium iridis (Ditmar) Fr.

DESCRIPTION: **Sporangia** stalked, white, 1.5 mm tall. **Sporotheca** erect, globose or slightly depressed, 0.4 to 0.6 mm in diameter. **Stalk** slender, cylindrical, usually two-thirds or more of the total height, yellow or yellowish brown, translucent. **Hypothallus** inconspicuous. **Peridium** thin, membranous, nearly or quite colorless, more or less densely covered with white lime crystals. **Columella** white or nearly so, turbinate, depressed-globose or subglobose. **Capillitium** delicate, consisting of colorless or pale yellow-brown branching threads, always hyaline at the tips. **Spores** brown in mass, pale violaceous by transmitted light, fairly warted or nearly smooth, 7 to 9 µm in diameter. **Plasmodium** yellowish or brown.

SUBSTRATE: Dead leaves and other types of plant debris.

COMMENTS: *Didymium nigripes* (Link) Fr. [page 159, Figure 8-37], a closely related species, has a dark brown columella and a brown to blackish brown stalk. *Didymium proximum* Berk. & M.A. Curtis is very similar to *D. iridis*; the two species are best distinguished by the color of their plasmodia, which is yellow in *D. proximum* and cream or brown in *D. iridis*. In addition, the sporangia of *D. proximum* tend to be somewhat taller than wide and the columella is usually yellow, not white. Interestingly, the trophic stages (i.e., amoeboflagellate cells and plasmodia) of *D. proximum* appear to be common in some forest soils.

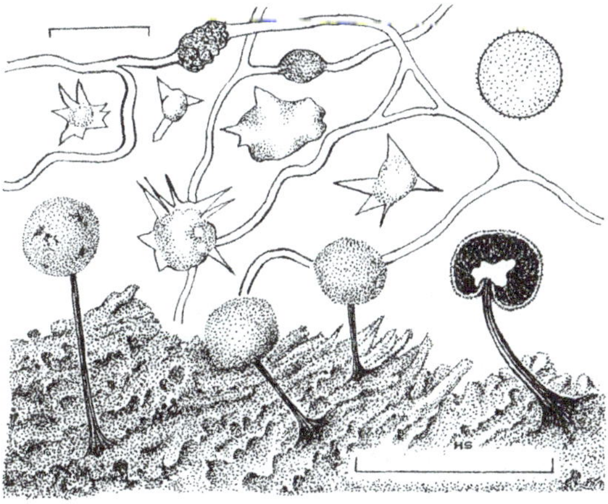

Figure 7-20B. *Didymium iridis*. Several sporangia, one sectioned to show the columella (lower portion of the drawing). A single spore, lime crystals, and portions of the capillitium (upper portion of drawing).

Figure 7-21A. *Didymium melanospermum* (©Yuri Novozhilov)

Didymium melanospermum (Pers.) T. Macbr.

DESCRIPTION: **Sporangia** stalked (or sometimes sessile), gregarious, up to 1 mm tall. **Sporotheca** subglobose or depressed, deeply umbilicate below, 0.5 to 1.0 mm in diameter. **Stalk** usually short, stout, rarely up to two-thirds the total height, dull brown or black, often completely immersed in the umbilicate base of the sporangium. **Hypothallus** dark but inconspicuous. **Peridium** firm, dull brown or black, more or less densely covered with white lime crystals. **Columella** prominent, hemispherical, dark or pallid, calcareous. **Capillitium** consisting of sparingly branched, pale to purple-brown, sinuous threads varying from slender to robust, often bearing dark, nodular thickenings. **Spores** black in mass, dull purplish brown by transmitted light, strongly warted or spinulose, 10 to 14 µm in diameter. **Plasmodium** colorless or dull gray.

SUBSTRATE: Decaying wood, twigs, and dead leaves.

COMMENTS: The distinguishing characteristics of this species are the robust, deeply umbilicate sporangia on short stalks. *Didymium minus* (Lister) Morgan, a closely related species, has smaller (0.4 to 0.6 mm in diameter) sporangia, relatively longer stalks, and smaller (8 to 11 µm in diameter), less strongly marked spores. *Didymium clavus* (Alb. & Schwein.) Rabenh. [page 158, Figure 8-34] has distinctive discoid, plate-shaped sporangia that are similar in appearance to those of *Diderma hemisphaericum* (Bull.) Hornem., from which it can be easily distinguished by its crystalline lime and a peridium consisting of only a single layer. The sporotheca of *Didymium squamulosum* (Alb. & Schwein) Fr. & Palmquist [page 159, Figure 8-38] is globose to subglobose and has a white, limy stalk. Since the stalks of *D. melanospermum*, *D. minus*, and *D. clavus* are dark, the color of the stalk of *D. squamulosum* is an important diagnostic feature.

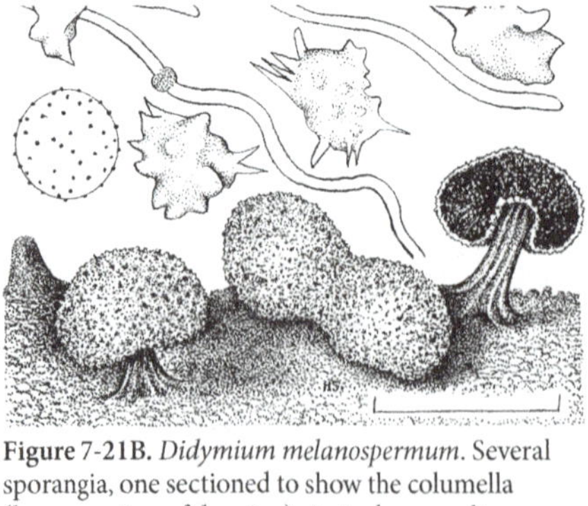

Figure 7-21B. *Didymium melanospermum*. Several sporangia, one sectioned to show the columella (lower portion of drawing). A single spore, line crystals, and portions of the capillitium (upper portion of drawing).

Figure 7-22A. *Echinostelium minutum* (©Yuri Novozhilov)

Echinostelium minutum de Bary

DESCRIPTION: **Sporangia** stalked, scattered to gregarious, 0.3 to 0.5 mm tall. **Sporotheca** erect or nodding, globose, 40 to 50 µm in diameter and 0.3 to 0.5 mm tall. **Stalk** translucent, colorless or pale yellow, 0.2 to 0.4 mm long. **Hypothallus** inconspicuous. **Peridium** not persisting in mature sporangia except (occasionally) as a collar at the base of the sporotheca, spore mass pale pink or white before dispersal. **Columella** slender, short, not exceeding 10 µm in total height. **Capillitium** scanty, branched with hooked free ends. **Spores** pink or white in mass, colorless by transmitted light, smooth but with thickened, more or less circular, plate-like areas evident in the spore wall, 7 to 8 µm in diameter. **Plasmodium** colorless.

SUBSTRATE: Tree bark placed in moist chamber cultures.

COMMENTS: This is far and away the most widespread and abundant species of *Echinostelium*. *Echinostelium cribrarioides* Alexop. is slightly larger and has a capillitium that forms a complete network with few or no free ends. Descriptions and a key to some of the other species of *Echinostelium* are provided by Whitney (1980) and Adamonyte (2006).

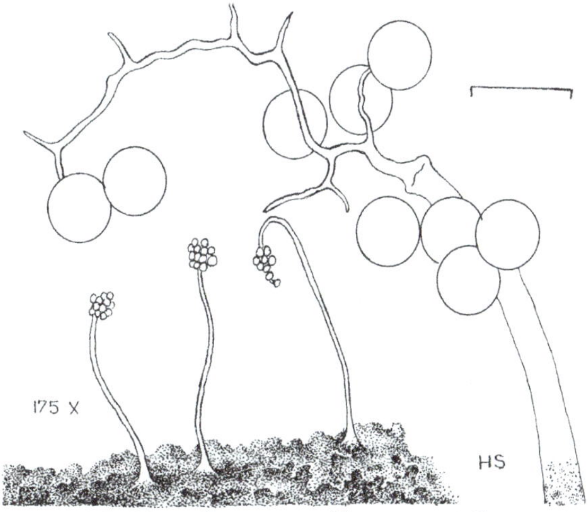

Figure 7-22B. *Echinostelium minutum*. Three sporangia (lower portion of drawing). A highly magnified view of the capillitium and sprores (upper portion of drawing).

Figure 7-23A. *Enerthenema papillatum* (©Kim Fleming)

Enerthenema papillatum (Pers.) Rostaf.

DESCRIPTION: **Sporangia** stalked, scattered, 1.0 to 1.5 mm tall. **Sporotheca** globose, 0.4 to 0.7 mm in diameter. **Stalk** black, opaque, attenuate above, about half the total height of the sporangium. **Hypothallus** membraneous, discoid or common to a group of sporangia. **Peridium** not persisting in mature sporangia, spore mass at first fuscous but becoming purplish brown or ferruginous as the spores are dispersed. **Columella** extending to the apex of the sporangium and there expanded as a small, shining, cup- or funnel-shaped apical disk not exceeding 0.2 mm in diameter. **Capillitium** attached to the apical disk, the threads long, dark, flexuous, sparsely branched. **Spores** olivaceous fuscous in mass, grayish brown by transmitted light, minutely warted, 10 to 12 µm in diameter. **Plasmodium** watery white.

Figure 7-23B. *Enerthenema papillatum*, Several sporangia (lower portion of drawing). A single sporangium after most of the spores have been lost (upper right). A portion of one of the capillitial branches and a single spore (upper right).

SUBSTRATE: Decaying wood and bark; also appearing on tree bark (especially that of coniferous trees) in moist chamber cultures.

COMMENTS: The "May-pole" arrangement of columella and branches of the capillitium make this species easy to recognize. *Enerthenema melanospermum* T. Macbr. & G.W. Martin, an alpine species likely to be encountered in western North America, can be distinguished on the basis of its intensely black color and larger (11 to 14 µm in diameter), more strongly marked spores.

Figure 7-24A. *Fuligo septica* (©Steve Stephenson)

Fuligo septica (L.) F.H. Wigg.

DESCRIPTION: **Aethalia** solitary to gregarious (but usually not widely separated), pulvinate, usually relatively large, 2 to 20 cm in extent and 1 to 3 cm thick. **Hypothallus** white, consisting of several layers and forming a layer around the aethalium. **Cortex** white to pale or bright pink or red to bright yellow, usually fairly thick, calcareous, fragile. **Capillitium** physaroid, consisting of white, yellow or reddish fusiform lime nodes connected by hyaline threads, sometimes scanty. **Spores** dark gray or dull black in mass, light purplish brown by transmitted light, nearly smooth to minutely spinulose, 6 to 9 μm in diameter. **Plasmodium** yellow.

SUBSTRATE: Decaying wood and bark, forest floor litter, wood debris and soil; sometimes fruiting on living plants.

COMMENTS: This is one of the most common, conspicuous, and best known myxomycetes. Several other species of *Fuligo* are occasionally encountered, but all of these have spores greater than 10 μm in diameter, and this feature alone distinguishes them from *F. septica*. *Fuligo intermedia* T. Macbr. has a thin yellowish gray or brown cortex and globose spores 10 to 12 μm in diameter, *F. cinerea* (Schwein.) Morgan has a white cortex and elliptical spores typically 13 to 15 μm in diameter, and *F. megaspora* Sturgis is characterized by an extremely thick, spongy cortex and very large (15 to 20 μm in diameter), strongly marked spores.

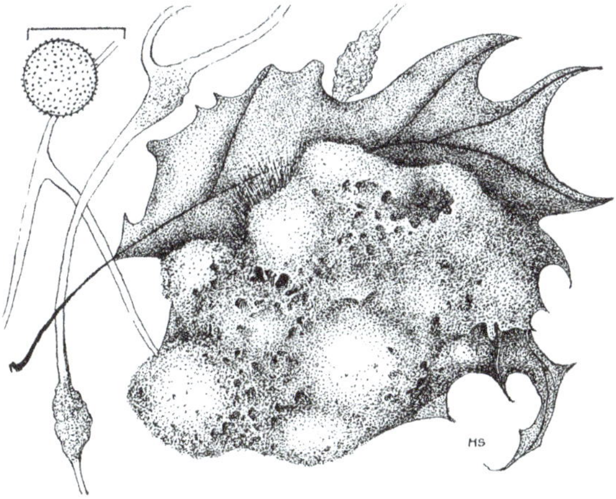

Figure 7-24B. *Fuligo septica* A single aethalium on oak leaves at about actual size (right side of drawing). A single spore and portions of the capillitium (left side of drawing).

Figure 7-25A. *Hemitrichia calyculata* (©Jeffery Gruber)

Hemitrichia calyculata (Speg.) M.L. Farr

DESCRIPTION: **Sporangia** stalked, scattered to gregarious, 1 to 3 mm tall. **Sporotheca** turbinate to subglobose, erect, 0.8 to 1.2 mm in diameter. **Stalk** slender, reddish brown to black, 0.5 to 2.0 mm long. **Hypothallus** irregular to discoid, light to dark reddish brown. **Peridium** membranous, bright to dark yellow, persisting in the lower half as a distinct calyculus that often has a petaloid margin. **Capillitium** consisting of a dense network of smooth yellow threads, marked with four to five spiral bands, 5 to 7 µm in diameter. **Spores** dull yellow in mass, pale yellow by transmitted light, 7 to 8 µm in diameter, minutely spiny to delicately reticulate. **Plasmodium** yellow but becoming red during the process of fruiting.

SUBSTRATE: Decaying wood and (less commonly) bark.

COMMENTS: This species, which is listed as *Hemitrichia stipitata* (Massee) T. Macbr. in some older monographs, is easily confused with *H. clavata* (Pers.) Rostaf. [page 160, Figure 8-41]. The latter is distinguished from *H. calyculata* on the basis of the much deeper, vase-like calyculus which gradually tapers downward to merge into the short, thick stalk. The capillitium of *H. clavata* also tends to be rougher than that of *H. calyculata*, and the sporangia produced by the former are more likely to be crowded closely together. *Trichia decipiens* (Pers.) T. Macbr. [page 141] is similar in general appearance to *H. clavata* but can be distinguished by its very different capillitium, which consists of free elaters with elongated tips. *Hemitrichia montana* (Morgan) T. Macbr., a snowbank species known only from alpine areas, has larger (10 to 13 µm in diameter) spores.

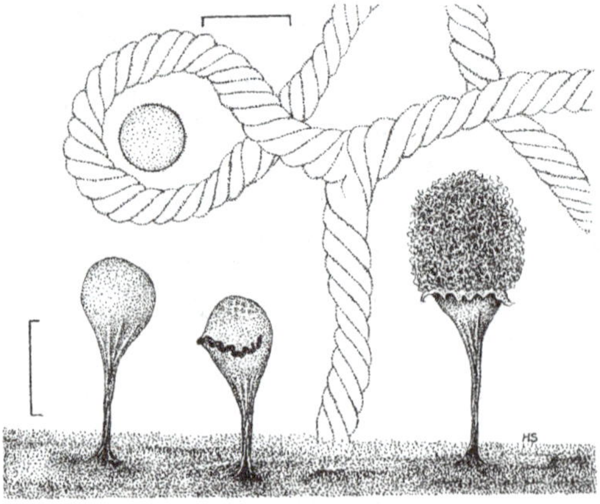

Figure 7-25B. *Hemitrichia calyculata*. Three sporangia (lower portion of drawing). One spore and a portion of the capillitium (upper portion of drawing).

Hemitrichia pardina (Minakata) Ing

Figure 7-26A. *Hemitrichia pardina* (©Alison Pollack)

DESCRIPTION: **Sporangia** stalked (more rarely sessile), scattered, 0.4 to 0.8 mm tall. **Sporotheca** erect, globose, 0.2 to 0.4 mm in diameter. **Stalk** usually about half the total height of the fruiting body but sometimes longer, cylindrical, black. **Hypothallus** inconspicuous. **Peridium** persistent, dull olive yellow with conspicuous wart-like structures present. **Capillitium** consisting of long, sparingly branched and anastomosing threads, 2 to 4 µm in diameter, with a few faint spirals and covered with short spines. **Spores** olive-yellow in mass, pale greenish yellow by transmitted light, faintly warted, 9 to 11 µm in diameter. **Plasmodium** white.

SUBSTRATE: Appearing in moist chamber culture on bark from living trees and dead leaves.

COMMENTS: Because of the dark, wart-like structures on the peridium, *Hemitrichia pardina* is unlikely to be confused with any other species. The most morphologically similar species is *H. minor* G. Lister. In fact, *H. pardina* was once considered to be a variety of *H. minor*. However, the consistent presence of the wart-like structures on the peridium of the former is the major distinguishing feature for the two species. *Perichaena pedata* (Lister & G. Lister) G. Lister ex E. Jahn and *Trichia subfusca* Rex might be confused with either of these two species of *Hemitrichia*, but both have a distinctly different capillitium.

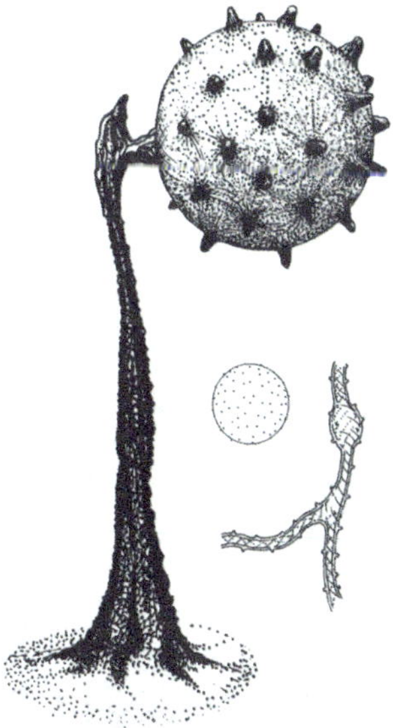

Figure 7-26B. *Hemitrichia pardina*. One sporangium (left side of drawing). One spore and a portion of the capillitium (right side of drawing).

Figure 7-27A. *Hemitrichia serpula* (©Jonathan Mays)

Hemitrichia serpula (Scop.) Rostaf. ex Lister

DESCRIPTION: Plasmodiocarps terete, 0.4 to 0.6 µm in diameter, branching freely and usually forming a definite reticulum that is often several cm in extent. **Hypothallus** usually evident, similar in color to the peridium. **Peridium** bright yellow to rusty or brownish, persistent, membranous, transparent, somewhat thicker below, splitting longitudinally to expose the spore mass. **Capillitium** a tangled mass of sparingly branched yellow threads, spinulose, 4 to 6 mm in diameter, with spiny free tips. **Spores** golden yellow in mass, pale yellow by transmitted light, coarsely reticulate, 10 to 16 µm in diameter. **Plasmodium** at first white and then becoming yellow.

SUBSTRATE: Decaying wood and wood debris.

COMMENTS: This is one of the most distinctive myxomycetes and is not likely to be confused with any other species. *Cornuvia serpula* (Wigand) Rostaf., *Dianema corticatum* Lister, *Perichaena chrysosperma* (Curr.) Lister [page 164, Figure 8-53], and *P. vermicularis* (Schwein.) Rostaf. also characteristically produce plasmodiocarps, although nothing quite so elaborate as the highly branched reticulum of *Hemitrichia serpula*. Instead, the plasmodiocarps of these four other species tend to be sparingly branched, often consisting of one to several slender, elongated worm-like or donut-shaped structures. *Dianema corticatum*, which appears to be predominantly alpine in distribution, has spores that occur in clusters of two to six. *Perichaena chrysosperma* and *P. vermicularis* are sometimes difficult to distinguish, but the former has somewhat smaller spores than the latter (8 to 10 µm versus 10 to 14 µm in diameter) and a spiny capillitium (the capillitium of *P. vermicularis* is usually nearly spineless). *Perichaena chrysosperma* tends to produce fruitings that are more sporangiate than plasmodiocarpous (fruitings of *P. vermicularis* are typically plasmodiocarps), and some fruiting of *P. chrysosperma* have a distinctive donut shape. *Cornuvia serpula* has a distinctive capillitium marked with prominent ring-like thickenings.

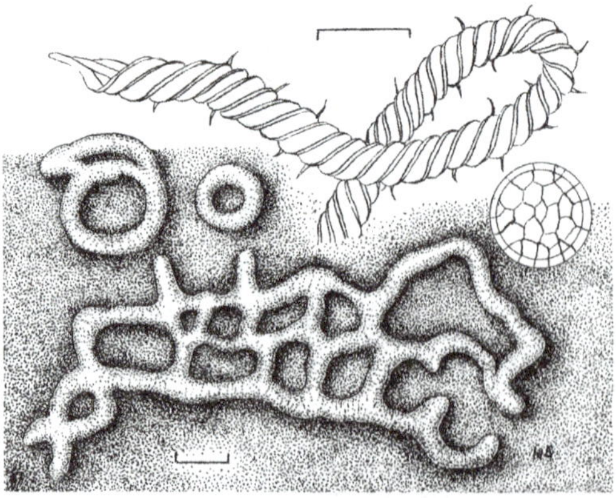

Figure 7-27B. *Hemitrichia serpula*. Three plasmodiocarps of different sizes (lower portion of drawing). A single spore and a portion of the capillitium (upper portion of the drawing).

Figure 7-28A. *Lamproderma arcyrioides* (©Alain Michaud)

Lamproderma arcyrioides (Sommerf.) Rostaf.

DESCRIPTION: **Sporangia** stalked or rarely sessile, scattered to gregarious, up to 1.5 mm tall. **Sporotheca** subglobose with a somewhat flattened base, 0.4 to 0.9 mm in diameter. **Stalk** black, up to three-fourths of the total height. **Hypothallus** membranous, reddish-brown. **Peridium** membranous, shining, iridescent blue, often with violet or bronze reflections. **Columella** cylindrical, obtuse, reaching to two-thirds of the height of the sporotheca. **Capillitium** consisting of pale threads, usually pale brown or darker and then only in the middle, arising from the upper part of the columella and branched to form a dense reticulum, very slender towards the tips. **Spores** black in mass, purplish gray by transmitted light, minutely spinulose, 8 to 11 µm in diameter. **Plasmodium** white.

SUBSTRATE: Various types of plant debris and sometimes living plants in alpine and lowland habitats, in the former occurring near the edge of melting snowbanks.

COMMENTS: This species occurs in both alpine and lowland habitats but is relatively more common in alpine habitats. In alpine habitats, it can be distinguished from most other species of *Lamproderma* by its smaller spores. The other more frequently encountered species of this genus in lowland habitats typically have a much longer stalk and a quite different capillitium.

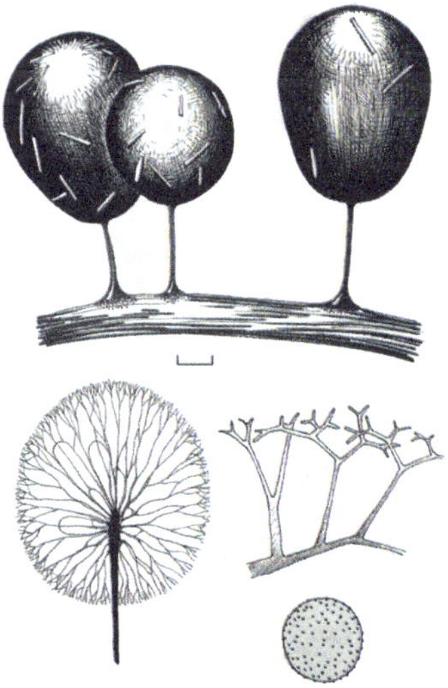

Figure 7-28B. *Lamproderma arcyriodes*. Three sporangia (upper portion of drawing). A semidiagrammatic sketch of a single sporangium showing the relationship of the capillitium and columella (lower left portion of the drawing). A portion of the capillitium and a single spore (lower right portion of the drawing). Scale bar = 0.1 mm.

Figure 7-29A. *Lamproderma columbinum* (©Alain Michaud)

Lamproderma columbinum (Pers.) Rostaf.

DESCRIPTION: **Sporangia** stalked, scattered or in small groups, 2.0 to 4.0 mm tall. **Sporotheca** erect, globose or ellipsoid, 0.5 to 1.0 mm in diameter. **Stalk** relatively long, stout, usually at least two-thirds of the total height, nearly cylindrical, black. **Hypothallus** dark purple. **Peridium** membranous, persistent, violet or purple and iridescent. **Columella** cylindrical with a blunt apex, reaching about the center of the sporotheca. **Capillitium** brownish purple, arising from the entire length of the columella, rigid forking near the columella and then anastomosed to form an open net with large meshes. **Spores** black in mass, smoky brown by transmitted light, minutely warted, 10 to 13 μm in diameter. **Plasmodium** white.

SUBSTRATE: Commonly occurring in association with bryophytes on moist decaying logs, on moss-covered rocks in shaded ravines, and on bryophyte-covered soil.

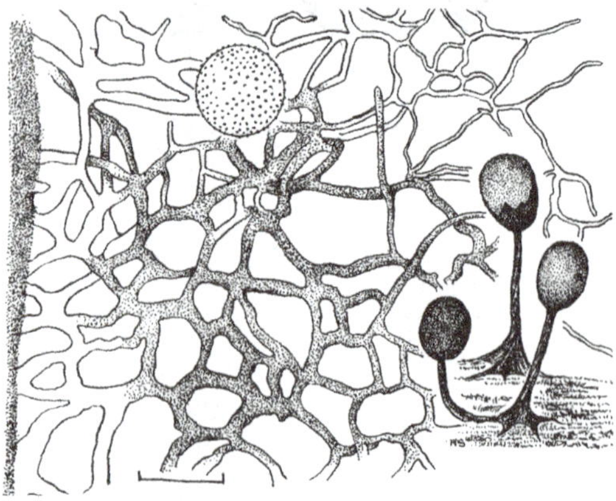

Figure 7-29B. *Lamproderma columbinum*. A group of sporangia (lower right). One spore and a semidiagrammatic sketch showing a single portion and a portion of the capillitium (left side of drawing).

COMMENTS: *Lamproderma columbinum* is the most robust non-alpine species in the genus likely to be encountered, although it is more common in some regions (e.g., the western United States) than others. The overall size, rigid capillitium, and its association with bryophytes are the major distinguishing features. The portion of the peridium at the base of the sporotheca of *L. columbinum* tends to persist after the upper portion has been lost. *Lamproderma arcyrioides* is morphologically similar to *L. columbinum* and can occur in lowland habitats, but this species is much smaller, and less robust and is not typically associated with bryophytes.

Figure 7-30A. *Lamproderma sauteri* (©Alain Michaud)

Lamproderma sauteri Rostaf.

DESCRIPTION: **Sporangia** stalked (or sometimes sessile), 1 to 4 mm tall. **Sporotheca** globose or slightly depressed, erect, 1 to 2 mm in diameter. **Stalk** usually short, rarely exceeding the height of the sporangium, black. **Hypothallus** firm, well developed. **Peridium** membranous, persistent, dark blue with a metallic luster. **Columella** reaching approximately the center of the sporangium, truncate or slightly enlarged at the tip. **Capillitium** coarse, rigid, purplish brown, with paler tips, appearing hoary after the spores have been dispersed. **Spores** black in mass, purplish brown by transmitted light, verrucose or with blunt spines, 12 to 15 µm in diameter. **Plasmodium** colorless or white.

SUBSTRATE: Decaying wood, wood debris, and dead leaves (usually of coniferous trees); occasionally fruiting on low shrubs and other living plants.

COMMENTS: This is the most common snowbank species of *Lamproderma* found in alpine regions of the temperate zone. *Meriderma carestiae* (Ces. &. De Not.) Mar. Mey. & Poulain [page 162, Figure 8-48], which is listed as *Lamproderma carestiae* (Ces. & De Not.) Meyl. in older texts, is very similar but has somewhat smaller (10 to 12 µm in diameter) spores and distinctive funnel-shaped tips on the ultimate branches of the capillitium. *Meriderma cribrarioides* (F.) Mar. May. & Poulain [page 163, Figure 8-49] is another morphologically similar species in which the capillitium has the same distinctive feature. The two species of *Meriderma* can be distinguished on the basis of their spores, with those of *M. cribrarioides* being conspicuously banded reticulate. *Lamproderma arcyrioides* (Summerf.) Rostaf., another very common alpine species, differs from *M. carestiae, M. cribrarioides,* and *L. sauteri* by its smaller (8 to 11 µm in diameter), less strongly marked spores and a more delicate capillitium. *Comatricha fusiformis* (Kowalski) Kowalski (listed as *L. fusiforme* Kowalski in earlier texts) has distinctive fusiform sporangia and a membranous silvery gray peridium.

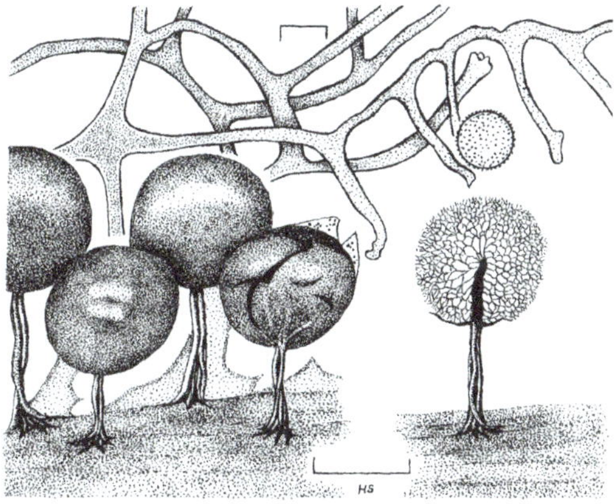

Figure 7-30B. *Lamproderma sauteri*. A group of sporangia (lower left). A semidiagrammatic sketch of a single sporangium showing the relationship of the columbella to the columella after dehiscence (lower right). A single spore and portions of the capillitium (upper portion of drawing).

Figure 7-31A. *Lamproderma scintillans* (©Joe Takano)

Lamproderma scintillans (Berk. & Broome) Morgan

DESCRIPTION: **Sporangia** stalked, scattered, 1 to 2 mm tall. **Sporotheca** erect, globose, erect, 0.3 to 0.4 mm in diameter. **Stalk** relatively long, usually at least two-thirds of the total height, slender, nearly cylindrical, black or dark brown. **Hypothallus** discoid, inconspicuous. **Peridium** persistent, metallic silvery, blue, or bronze, iridescent. **Columella** cylindrical, truncate, not exceeding the center of the spore mass. **Capillitium** dense, consisting of rigid, straight, sparingly branched and anastomosing threads with numerous free ends, branches of the capillitium pallid or colorless as they leave the columella, elsewhere brown. **Spores** brown in mass, violet-gray by transmitted light, regularly and distinctly warted, 7 to 9 μm in diameter. **Plasmodium** watery white.

SUBSTRATE: Dead leaves; less commonly on wood, dung, and other types of plant debris.

COMMENTS: *Lamproderma scintillans* is the non-alpine species of *Lamproderma* most likely to be encountered by the average collector, but the similar *Collaria arcyrionema* (Rostaf.) Nann.-Bremek. (listed as *Lamproderma arcyrionema* Rostaf. in earlier texts) [page 152, Figure 8-17] is more common, especially in the field (most records of *L. scintillans* are from moist chamber cultures). The two species can be distinguished by their spores and capillitium; *L. scintillans* has distinctly warted spores and a capillitium consisting primarily of rigid, straight branches, while *C. arcyrionema* has minutely punctate spores and a capillitium that is flexuous and highly branched. Moreover, the usual substrate for *L. arcyrionema* is decaying wood. *Lamproderma columbinum* (Pers.) Rostaf. [page 110], a species generally restricted to high-elevation coniferous forests, has iridescent violet or purple sporangia and spores 10 to 13 µm in diameter. *Macbrideola scintillans* H.C. Gilbert resembles *L. scintillans* but is much smaller (less than 0.3 mm tall) and has a hollow, tubular stalk. The latter is an important distinguishing characteristic of members of the genus *Macbrideola* and can be determined by looking for air bubbles that become trapped in the interior of the stalk when the specimen is mounted on a slide. In *Collaria*, *Comatricha*, and *Lamproderma* the stalk is fibrous or solid. *Macbrideola cornea* (G. Lister & Cran) Alexop. [page 162, Figure 8-47] is somewhat similar to *M. scintillans* but lacks a silvery peridium. *Macbrideola scintillans* and *M. cornea*, as is the case for all other species in the genus, are known almost exclusively from fruitings appearing on bark in moist chamber cultures.

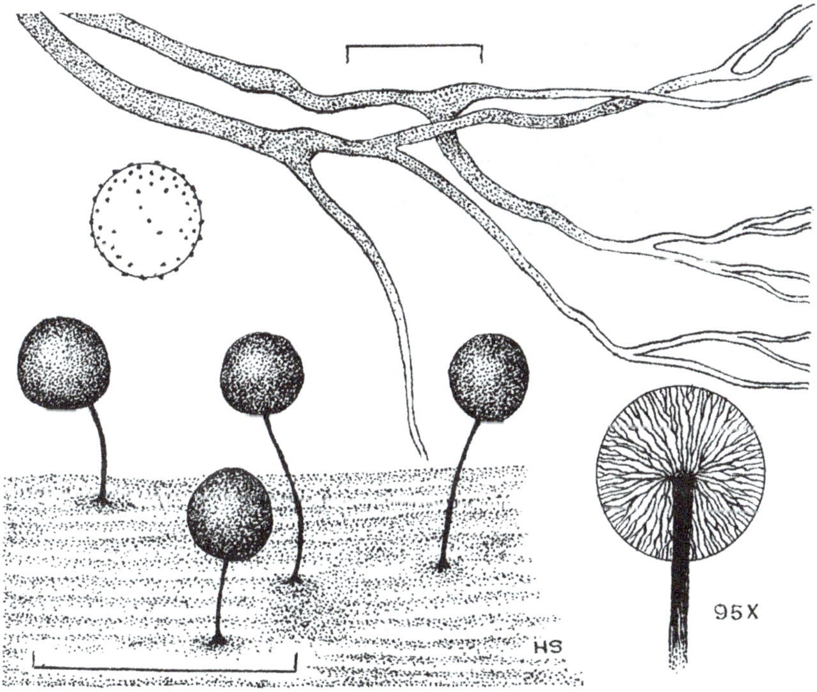

Figure 7-31B. *Lamproderma scintillans*. Several sporangia (lower left). A semidiagrammatic sketch of a single sporangium showing the relationship of the capillitium to the columella (lower right). A single spore and a portion of the capillitium (upper portion of drawing).

Figure 7-32A. *Leocarpus fragilis* (©Henri Koskinen)

Leocarpus fragilis (Dicks.) Rostaf.

DESCRIPTION: **Sporangia** stalked or nearly sessile, gregarious or clustered, 2 to 4 mm tall. **Sporotheca** short-cylindrical to obovate to subglobose, 0.6 to 1.6 mm in diameter. **Stalk** weak, often flattened, pale yellow or ochraceous, essentially representing an extension of the membranous hypothallus, 0.4 to 2.5 mm long. **Hypothallus** membranous, pale ochraceous. **Peridium** pale yellow to ochraceous to chestnut brown or deep maroon, smooth, shining, brittle, consisting of three layers, the outer layer cartilaginous, the middle layer calcareous, and the inner layer membranous, hyaline. **Capillitium** of two types, the first consisting of a network of rigid, branching, white calcareous nodes and the second consisting of a network of slender, colorless, flattened tubules usually expanded at the junctions. **Spores** black in mass, brown by transmitted light, with a paler area at one side, coarsely warted, 12 to 14 µm in diameter. **Plasmodium** yellow.

SUBSTRATE: Forest floor litter and wood debris; sometimes fruiting on living plants.

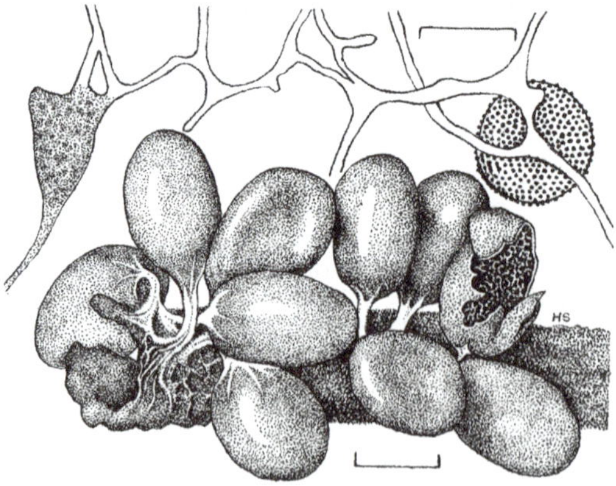

Figure 7-32B. *Leocarpus fragilis*. Clustered sporangia on a twig (lower portion of drawing). A single spore and a portion of the capillitium (upper portion of the drawing).

COMMENTS: This is another species not likely to be confused with any other myxomycete, although a small fruiting could be mistaken for a mass of insect eggs. *Leocarpus fragilis* is more likely to be encountered in coniferous forests than in broadleaf forests and appears to be confined almost exclusively to temperate and cool temperate regions of the world.

Figure 7-33A. *Lepidoderma tigrinum* (©Marius Muessler)

Lepidoderma tigrinum (Schrad.) Rostaf.

DESCRIPTION: **Sporangia** stalked (or rarely sessile), gregarious or scattered, up to 2.8 mm tall. **Sporotheca** subglobose, 0.8 to 1.5 mm in diameter. **Stalk** stout, furrowed, bright orange-brown, spongy, with orange lime crystals inside, 1.0 to 2.0 mm long. **Hypothallus** dull yellow or orange-brown **Peridium** cartilaginous, olive or purplish gray, the surface incompletely covered by rounded to angular plates of crystalline lime, opaque, dark gray or dull orange-brown. **Columella** large, hemispherical, similar in structure to the stalk and containing the same type of lime crystals. **Capillitium** profuse, consisting of dark, sparingly branched, straight or flexuous threads radiating from the columella. **Spores** black in mass, dull grayish brown by transmitted light, minutely spinulose, 10 to 13 µm in diameter. **Plasmodium** orange-yellow.

SUBSTRATE: Decaying coniferous wood; usually wood with leafy liverworts present.

COMMENTS: This distinctive species is not likely to be confused with any other myxomycete. It typically fruits late in the season. *Lepidoderma chailletii* Rostaf. [page 160, Figure 8-42], *L. carestianum* (Rabenh.) Rostaf., and *L. granuliferum* (W. Phillips) R. E. Fr. are found in alpine areas. All three produce sessile fruiting bodies and thus can be distinguished from *L. tigrinum* by this character alone. *Lepidoderma carestianum* is typically plasmodiocarpous, although sporangiate sporangia are sometimes encountered. *Lepidoderma granuliferum* is exclusively plasmodiocarpous, while *L. chailletii* is sporangiate. *Lepidoderma granuliferum* has larger spores (15 to 20 µm in diameter) than those of *L. carestianum* (which are mostly less than 15 µm in diameter). Sporangiate forms of *L. carestianum* can be distinguished from *L. chailletii* by their smaller size (sporangia 0.5 to 1.5 mm in diameter versus 1.5 to 3.0 mm in diameter) and different spore ornamentation (densely spinulose spores versus spores with widely scattered spines). A key to most of the more common species of the genus *Lepidoderma* was provided by Kowalski (1971). As a result of recent molecular data, *Lepidoderma tigrinum* is no longer considered to be a member of the genus *Lepidoderma* and is now recognized as *Diderma tigrinum* (Schrad.) Prikhodko, Shchepin, Novozh., López-Vill., G. Moreno & Schnittler.

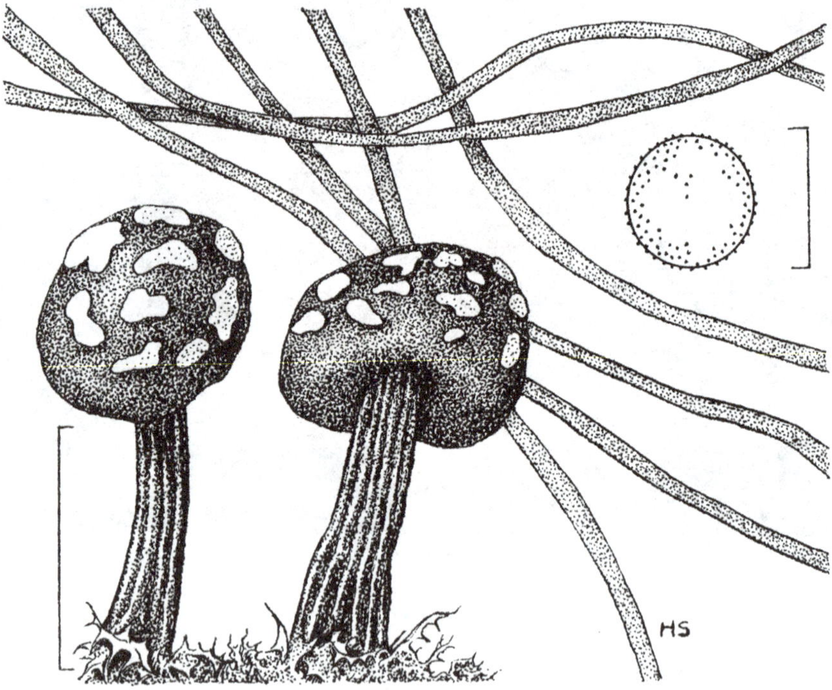

Figure 7-33B. *Lepidoderma tigrinum*. Two sporangia (lower portion of drawing). Portions of the capillitium and a single spore (upper portion of drawing).

Licea biforis Morgan

Figure 7-34. *Licea biforis* (©Steve Young)

DESCRIPTION: **Sporangia** sessile (or sometimes plasmodiocarpous), scattered to gregarious, fusiform, somewhat compressed and occasionally branched, 0.1 to 0.3 mm wide and 0.3 to 0.8 mm long. **Hypothallus** inconspicuous. **Peridium** yellow-brown to dark brown or nearly black, dehiscence by means of a preformed longitudinal fissure. **Spores** nearly smooth to asperulate, yellow-brown to brown in mass, clear yellow to almost colorless by transmitted light, 9 to 12 μm in diameter. **Plasmodium** at first watery white and then becoming brown.

SUBSTRATE: Dead bark; also appearing on bark and leaf litter in moist chamber cultures.

COMMENTS: No other species of *Licea* has sessile, elongated (canoe-shaped) sporangia opening by a longitudinal slit. Dehiscence in *Licea succulenticola* Mosquera, Lado, Estrada & Beltrán-Tej. [page 156, Figure 8-44] also involves a longitudinal slit, but the sporangia are subglobose, the spores are yellowish brown, and the usual substrate is distinctly different. *Licea succulenticola* is associated with decaying portions of succulent plants such as cacti.

Licea kleistobolus G.W. Martin

Figure 7-35. *Licea kleistobolus* (©Alison Pollack)

DESCRIPTION: **Sporangia** sessile, scattered to gregarious, discoid to subglobose, 0.04 to 0.15 mm in diameter. **Hypothallus** inconspicuous. **Peridium** dark brown to black below and bright coppery brown above, dehiscence by means of a preformed operculum. **Spores** ochraceous in mass, very pale yellow by transmitted light, smooth, 9 to 13 μm in diameter. **Plasmodium** watery blackish brown.

SUBSTRATE: Appearing on bark in moist chamber cultures.

COMMENTS: The bright coppery color of the operculum makes this an easy species to recognize. *Licea parasitica* (Zukal) G.W. Martin is similar in shape and size but has sporangia that are uniformly dark brownish gray in color. The sporangia of *L. biforis* are fusiform, somewhat compressed, and dehisce by means of a longitudinal fissure instead of a preformed operculum. *Licea tenera* E. Jahn is predominantly coprophilous and has bright yellow spores. A key to most of the more common species of *Licea* was provided by Keller and Brooks (1977).

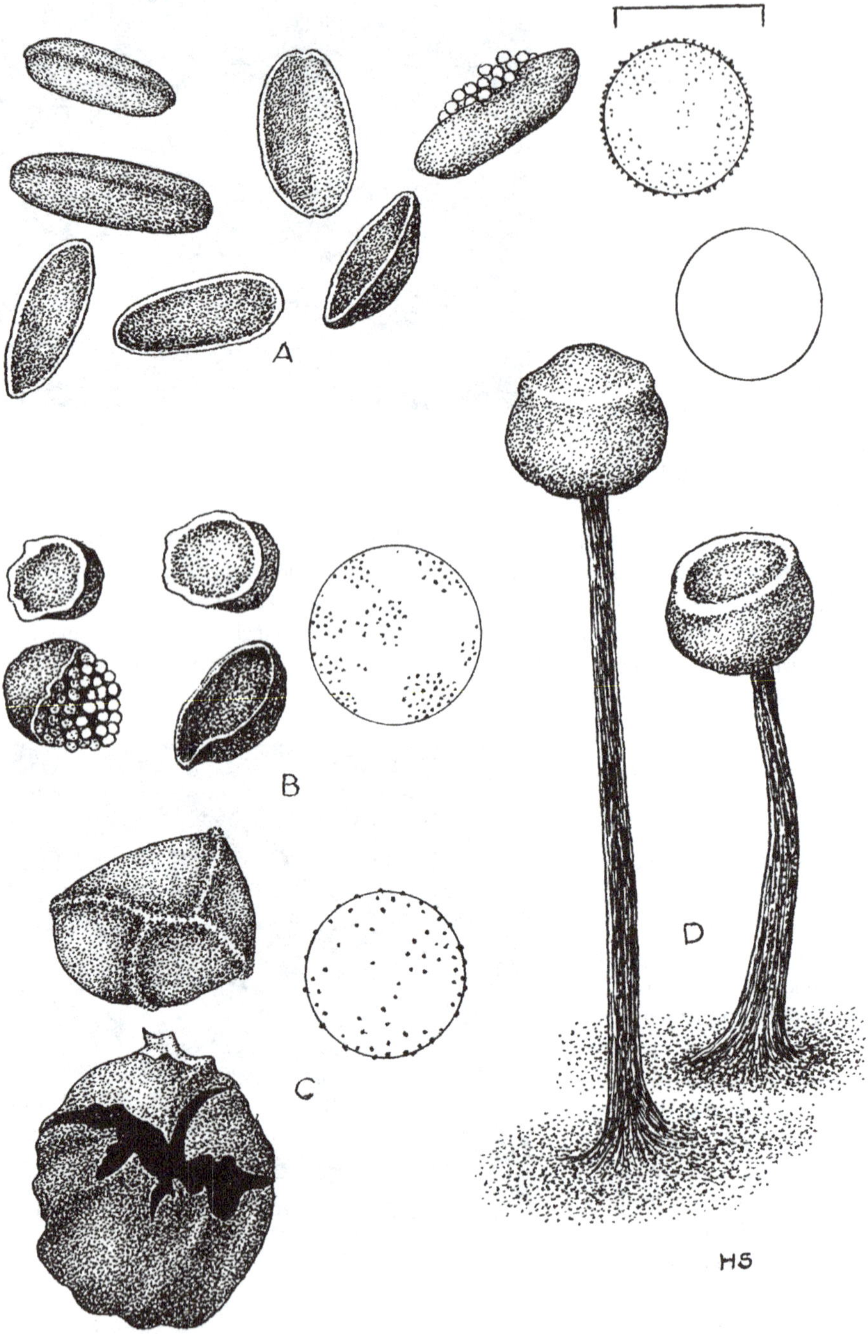

Figure 7-36. Fruiting bodies of four species of *Licea*. A. *Licea biforis*. B. *Licea kleistobolus*. C. *Licea minima*. D. *Licea operculata*.

Licea minima Fr.

Figure 7-37. *Licea minima* (©Alain Michaud)

DESCRIPTION: **Sporangia** sessile, scattered to gregarious, subglobose to pulvinate or angular, 0.2 to 0.4 mm in diameter. **Hypothallus** inconspicuous. **Peridium** umber or reddish brown to nearly black, opaque, with prominent ridges, these breaking apart into angular plates with dotted margins, plates becoming petaloid or widely reflexed in older specimens. **Spores** dark reddish brown in mass, smoky ferruginous to olivaceous brown by transmitted light, the spore wall thinner and paler on one side, minutely warted, 10 to 12 µm in diameter. **Plasmodium** at first watery drab or gray and then becoming yellowish or rosaceous brown.

SUBSTRATE: Decaying coniferous wood; also appearing on bark from coniferous trees in moist chamber cultures.

COMMENTS: This is one of the few species of *Licea* that can be spotted in the field. *Licea pusilla* Schrad. and *L. castanea* G. Lister are similar in appearance to *L. minima* but can be distinguished from the latter on the basis of spore characteristics; the spores of *L. pusilla* are larger (15 to 17 µm in diameter) than those of *Licea minima*, while the spores of *L. castanea* are smooth (the spores of *L. minima* are warted).

Licea operculata (Wingate) G.W. Martin

Figure 7-38. *Licea operculata* (©Alison Pollack)

DESCRIPTION: **Sporangia** stalked, scattered to gregarious, 0.4 to 1.0 mm tall. **Sporotheca** erect, urn-shaped or subglobose, 0.1 to 0.3 mm in diameter. **Stalk** subcylindrical, rough, usually two-thirds or more of the total height. **Hypothallus** inconspicuous. **Peridium** dull brown to almost black except for the yellow, iridescent operculum, dehiscence by means of a preformed operculum. **Spores** pale yellow in mass, colorless by transmitted light, smooth, 8 to 11 µm in diameter. **Plasmodium** dull orange.

SUBSTRATE: Appearing on bark in moist chamber cultures.

COMMENTS: *Licea pedicellata* (H.C. Gilbert) H.C. Gilbert, the only other relatively common species of *Licea* with which *L. operculata* might be confused, lacks an operculum and has a short, thick, furrowed stalk. Wrigley de Basanta and Lado (2005) provided a key to all of the stalked species of *Licea*.

Figure 7-39A. *Lycogala epidendrum* (©Anne-Marie Rantet Roux)

Lycogala epidendrum (L.) Fr.

DESCRIPTION: **Aethalia** usually several to many in a single fruiting, these scattered to crowded, subglobose to depressed-spherical or irregular from pressure, 3 to 15 mm broad. **Hypothallus** inconspicuous. **Cortex** (considered to be a peridium by some mycologists who work with myxomycetes) warted or merely roughened, pinkish gray or yellowish brown to deep olivaceous or nearly black, rather thin and fragile, especially above. **Pseudocapillitium** composed of long, branching and anastomosing flattened tubules marked with conspicuous transverse folds and wrinkles. **Spores** at first pink in mass, changing to pale ochraceous or pallid, colorless by transmitted light, reticulate, 6 to 8 µm in diameter. **Plasmodium** coral-red.

SUBSTRATE: Decaying wood and (less commonly) bark.

COMMENTS: This is one of the most widely distributed and best-known myxomycetes. Three other species of *Lycogala* are much less common. *Lycogala flavofuscum* (Ehrenb.) Rostaf. [page 162, Figure 8-46] produces generally larger (usually 2 to 4 cm broad) aethalia that have a much thicker, nearly smooth cortex. *Lycogala exiguum* Morgan and *L. conicum* Pers. [page 161, Figure 8-45] are both smaller than *L. epidendrum*. The former is darker and has a cortex thickly covered with dark purple-brown or black scales, while the aethalia produced by the latter are distinctively conical. Leontyev et al. (2022b), who carried out a detailed study of the peridium of *Lycogala epidendrum*, reported that their data indicated the presence of numerous previously unrecognized forms of this morphospecies that are likely to be described as species new to science. When a large series of specimens of this species collected worldwide was examined using molecular markers, the data suggested that the number of unrecognized forms is at least 60 (Leontyev et al. 2022a). As such, what has long been recognized as *L. epidendrum* actually consists of a very large species complex.

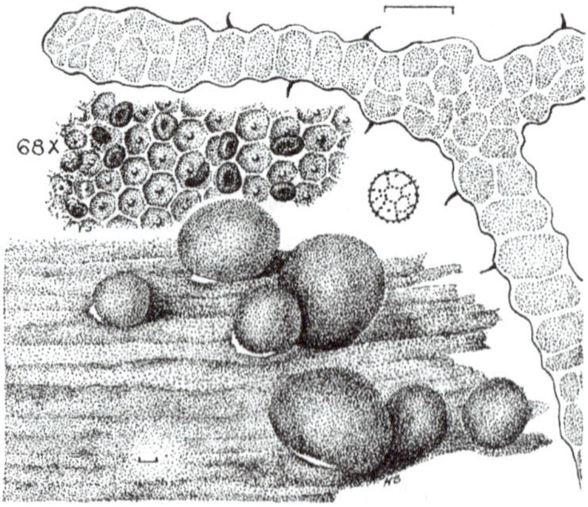

Figure 7-39B. *Lycogala epidendrum*. Group of aethalia (lower portion of drawing). Detailed view of the surface of the cortex (left of center). A single spore and a portion of the pseudocapillitium (upper right).

Figure 7-40A. *Metatrichia vesparia* (©Yuri Novozhilov)

Metatrichia vesparia (Batsch) Nann.-Bremek. ex G.W. Martin & Alexop.

DESCRIPTION: **Sporangia** stalked (or rarely sessile), gregarious to clustered, individual sporangia 1.0 to 4.5 mm tall. **Sporotheca** erect, obovate, usually firmly united into clusters, 0.5 to 1.0 mm in diameter. **Stalk** solid, rather thick when supporting several sporangia, brick red, 0.3 to 3.5 mm long. **Hypothallus** extensive, brownish red, often common to several sporangia. **Peridium** wine-red to dark maroon or sometimes nearly black, opaque, firm, shining, often with metallic reflections, dehiscence by a preformed dome-shaped operculum. **Capillitium** consisting of numerous long, free, rarely branched elaters, almost all of which are bent 180 degrees in the middle with the two halves coiled about one another, bearing three to five spiral bands and numerous spines 1 to 2 µm long, bright red to deep crimson, the tips blunt. **Spores** brownish red in mass, reddish orange by transmitted light, minutely warted, 9 to 11 µm in diameter. **Plasmodium** deep red to black.

SUBSTRATE: Decaying wood or bark; occasionally on dead leaves.

COMMENTS: The clustered sporangia, which resemble miniature paper wasp nests, are distinctive. *Metatrichia floriformis* (Schwein.) Nann.-Bremek. [page 163, Figure 8-50] is somewhat similar in both color and general structure but can be distinguished from *Metatrichia vesparia* by its smooth capillitium with elongated tips.

Figure 7-40B. *Metatrichia vesparia*. Two clusters of sporangia with fused stalks (lower portion of drawing). Note that in one sporangium the expanding capillitium is pushing the apical "cap." A single spore and a portion of the capillitium (upper left).

Figure 7-41A. *Mucilago crustacea* (©Gary Laursen)

Mucilago crustacea P. Micheli ex F.H. Wigg.

DESCRIPTION: **Aethalia** usually solitary, pulvinate, consisting of numerous anastomosing tubes with membranous walls, 1 to 7 cm long, 1 to 5 cm wide, and 1 to 2 cm thick. **Hypothallus** well developed, membranous to spongy, hyaline or white, often including dense masses of crystalline lime. **Cortex** pure white to creamy white or pale ochraceous, dense, spongy or flaky to pulverulent, composed of calcareous crystals. **Capillitium** consisting of a network of dark, branching and anastomosing threads, often with expanded portions. **Pseudocapillitium** composed of the walls of the constituent tubes, thin, hyaline, often iridescent. **Spores** black in mass, blackish brown or occasionally bright purplish brown by transmitted light, densely and unevenly verrucose or spinulose, 11 to 13 µm in diameter. **Plasmodium** creamy white to pale yellowish.

SUBSTRATE: Decaying wood, dead leaves, various other types of plant debris, and low-growing living plants.

Figure 7-41B. *Mucilago crustacea*. A single aethalium (lower left). A single spore (upper left). Portions of the capillitium and lime crystals (right side of drawing).

COMMENTS: The only other myxomycetes that might be confused with this species are *Fuligo cinerea* (Schwein.) Morgan, *F. intermedia* T. Macbr., and *F. septica* (L.) F.H. Wigg., which are discussed elsewhere in this book. The cortex of species of *Fuligo* is made up of calcareous granules, while the cortex of *Mucilago crustacea* consists of calcareous crystals.
On the basis of recent molecular data, the genus *Mucilago* is no longer valid, and *M. crustacea* is now recognized as *Didymium sponglosum* (Leyss.) J.M. García-Martín, J.C. Zamora & Lado.

Figure 7-42A. *Perichaena depressa* (©Laurie Leonard)

Perichaena depressa Lib.

DESCRIPTION: **Sporangia** sessile, usually crowded but occasionally scattered, depressed-pulvinate, polygonal by mutual contact, 0.1 to 1.5 mm in diameter. **Hypothallus** inconspicuous. **Peridium** chestnut to dark purplish brown or nearly black, consisting of two layers, the outer layer sometimes hoary or covered with amorphous or crystalline lime (calcium oxalate) deposits, closely appressed to the membranous inner layer, dehiscence circumscissile by means of a definite preformed operculum. **Capillitium** consisting of slender, simple or branched free elaters, yellow, 2 to 3 µm in diameter, minutely warted or spiny. **Spores** bright yellow in mass, pale yellow by transmitted light, minutely warted, 9 to 12 µm in diameter. **Plasmodium** colorless or yellowish.

SUBSTRATE: Decaying bark and (less commonly) wood; sometimes occurring on leaf litter and dung in moist chamber cultures.

COMMENTS: Typical fruitings of this species are easily recognized, but atypical fruitings may be mistaken for *Perichaena corticalis* (Batsch) Rostaf. [page 164, Figure 8-54]. The sporangia of the latter species, however, are not markedly depressed or angular. The sporangia of *Perichaena dictyonema* Rammeloo [page 165, Figure 8-55] are dome-shaped, scattered, and have the inner peridium prominently exposed around the middle. A key to all the common species of *Perichaena* was provided by Keller and Eliasson (1992).

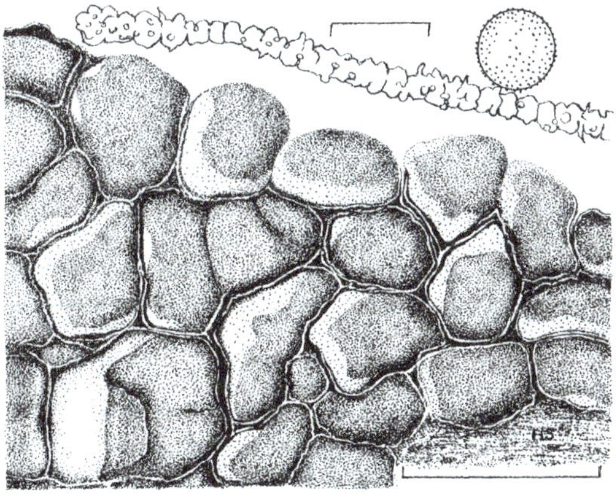

Figure 7-42B. *Perichaena depressa*. Crowded sporangia as seen from above (lower portion of drawing). A single spore and a portion of a strand (upper portion of drawing).

Figure 7-43A. *Physarella oblonga* (©Todd Elliott)

Physarella oblonga (Berk. & M.A. Curtis) Morgan

DESCRIPTION: **Sporangia** stalked or very rarely sessile and somewhat plasmodiocarpous, gregarious, up to 4 mm tall. **Sporotheca** erect or nodding, cylindrical, up to 1 mm in diameter. **Stalk** terete or flattened, hollow, red, translucent, 1.0 to 3.0 mm long. **Hypothallus** membranous, inconspicuous. **Peridium** yellow (or very rarely white), flecked with yellow, brown, or red (occasionally white) limy scales that may be sparse or form a nearly continuous limy crust, dehiscence apical, the outer portion of the peridium rupturing irregularly or in a more or less lobate fashion, the segments becoming recurved, exposing the spikes of the limy capillitium and leaving the spikeless inner peridial walls protruding as a cylindrical tubular pseudocolumella. **Capillitium** of two types, the first consisting of limy spike-like, simple or more or less branched processes arising from the inner walls of the outer portion of the peridium, the second consisting of a dense network of slender, violaceous threads bearing a few fusiform, yellow or white calcareous nodes. **Spores** grayish brown in mass, violaceous brown by transmitted light, punctate or asperulate to densely verruculose, 7 to 8 µm in diameter. **Plasmodium** yellow or white.

SUBSTRATE: Decaying wood and dead leaves.

COMMENTS: Typical sporangia of *Physarella oblonga*, which somewhat resemble miniature daffodils, are so distinctive that they are unlikely to be confused with those of any other myxomycete. The distribution of this species is primarily tropical, and it is rarely collected in temperate regions of the world.

Figure 7-43B. *Physarella oblonga*. Three sporangia (lower left). Note that the one on the left is still intact, while the other two have already dehisced. The margin of a plasmodiocarpous fruiting body (lower right). A portion of the capillitium and a single spore (upper portion of drawing).

Figure 7-44A. *Physarum albescens* (©Yuri Novozhilov)

Physarum albescens Ellis ex T. Macbr.

DESCRIPTION: **Sporangia** short to long stalked or sessile, densely gregarious, rarely scattered, up to 2.5 mm tall. **Sporotheca** subglobose to obovoid or forming short plasmodiocarps, commonly 0.7 to 1.5 mm in diameter. **Stalk** consisting of strands of the hypothallus, up to 2.0 mm long. **Hypothallus** conspicuous, veined, creamy yellow to pale orange. **Peridium** white, pale to brilliant yellow to dark orange or, when lime is lacking, dark blue, single but when the limy crust is especially thick, appearing double, membranous and iridescent when the lime is lacking. **Columella** absent, pseudocolumella rarely present. **Capillitium** dense, pale to dark brown forming a rigid, intricate reticulum, threads flattened, lime nodes sparse to numerous, large, round, elongated or branched, white, yellow, orange or red. **Spores** black to dark purple-brown in mass, purple to violet-brown by transmitted light, spinuose, 11-13 µm in diameter, often paler on one side. **Plasmodium** yellow or orange red.

SUBSTRATE: Various types of plant debris and sometimes living plants near melting snowbanks in alpine areas.

COMMENTS: *Physarum albescens* is a distinctive species unlikely to be confused with any other myxomycete commonly encountered in alpine snowbank habitats. It sometimes forms large fruitings that can be as much as a meter in total extent. Based on recent molecular data, *Physarum albescens* was placed in the genus *Badhamia* as *B. albescens* (Ellis ex T. Macbr.) J.M. García-Martín, J.C. Zamora & Lado.

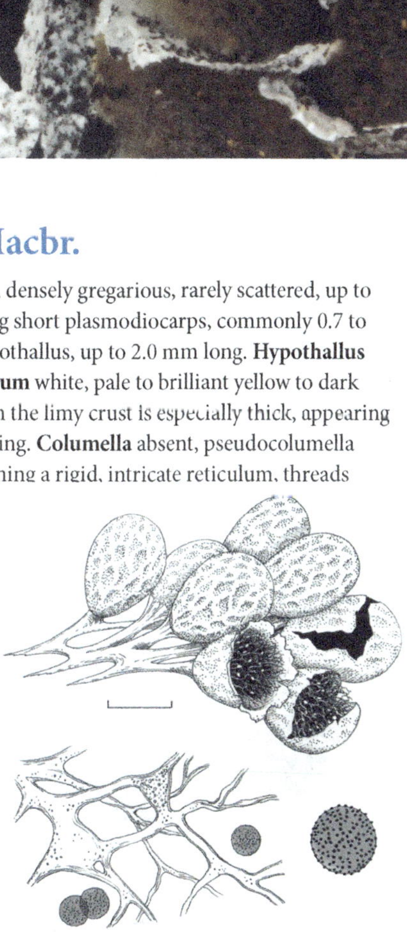

Figure 7-44B. *Physarum albescens.* A group of sporangia (upper portion of the drawing). A portion of the capillitium and three spores (lower left portion of the drawing). A single spore (lower right portion of the drawing).

Figure 7-45A. *Physarum album* (©Alain Michaud)

Physarum album (Bull.) Chevall.

DESCRIPTION: **Sporangia** stalked, gregarious, 1.0 to 1.5 mm tall. **Sporotheca** usually nodding, subglobose or lenticular, 0.3 to 0.7 mm in diameter. **Stalk** non-calcareous, slender, tapering, longitudinally wrinkled, and usually dark in color (becoming less so toward the apex), 0.7 to 1.2 mm long. **Hypothallus** inconspicuous. **Peridium** white or pale, encrusted with lime, the upper portion splitting into irregular fragments. **Capillitium** physaroid, consisting of a network of colorless slender threads with interspersed small, white lime nodes. **Spores** black in mass, pale lilaceous brown by transmitted light, minutely spinulose or nearly smooth, 8 to 10 µm in diameter. **Plasmodium** colorless or pale brown.

SUBSTRATE: Decaying wood or bark.

COMMENTS: Two very similar species are *Physarum viride* (Bull.) Pers. [page 168, Figure 8-65] and *P. flavicomum* Berk. The first, an exceedingly common species, is distinguished by the yellow (less commonly reddish orange) peridium and lime nodes. The second species, which is not as common as either *P. album* or *P. viride*, has an iridescent bronze or sometimes dusky yellow peridium that is often nearly limeless and usually persists only at the base of the sporotheca. *Physarum leucophaeum* Fr. & Palmquist closely resembles *P. album* but differs in having a shorter stalk (usually no more than half the total height) and more erect sporangia. Most of the capillitial threads of *P. album* originate from the base of the sporotheca, which is not the case for *P. leucophaeum*. Stalked sporangia of *P. notabile* T. Macbr., a very variable species in which a single fruiting may contain both stalked and sessile sporangia as well as short plasmodiocarps, resemble those of *P. leucophaeum*. Fruitings of *P. notabile*, however, usually are more extensive, the sporangia typically occur in tightly packed clusters, and the spores are much darker when viewed by transmitted light. Sporangia of *Physarum crateriforme* Petch [page 166, Figure 8-58], a species that typically appears on bark in moist chamber cultures, are usually stalked but can be sessile. The sporotheca is globose to somewhat elongated, but the distinctive feature is the presence of a prominent columella that is lacking in other morphologically similar species.

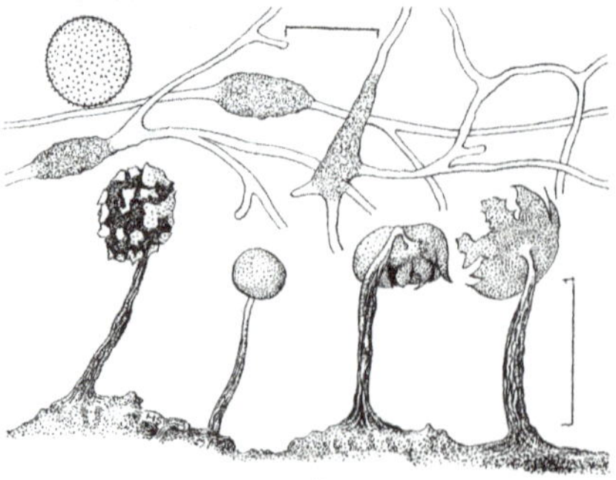

Figure 7-45B. *Physarum album*. Four sporangia showing different stages of dehiscence (lower portion of drawing). A single spores and portions of the capillitium (upper portion of drawing). Scale bar = 0.5 mm.

Figure 7-46A. *Physarum bivalve* (©Alain Michaud)

Physarum bivalve Pers.

DESCRIPTION: **Sporangia** plasmodiocarps or sessile, gregarious to crowded, laterally compressed, 0.5 mm to 1.0 cm or more long and 0.7 to 1.0 mm tall. **Hypothallus** inconspicuous. **Peridium** white to dark gray or brownish yellow, double, the outer layer usually thickly calcareous and white, the inner layer delicate, membranous, colorless, dehiscence by a more or less regular, preformed longitudinal fissure. **Capillitium** physaroid, abundant, consisting of numerous, large, angular, elongated or branching white lime nodes and short connecting filaments. **Spores** black in mass, dull violet brown by transmitted light, minutely and uniformly spinulose, 8 to 10 µm in diameter. **Plasmodium** gray, pallid or yellowish.

SUBSTRATE: Dead leaves and other plant debris on the forest floor; occasionally fruiting on living plants.

COMMENTS: The individual fruiting bodies of this species look like miniature clams turned up on edge. *Physarum echinosporum* Lister [page 167, Figure 8-61] is morphologically similar to *P. bivalve* but has an iridescent pale purple inner peridium while the inner peridium in *P. bivalve* is essentially colorless. Moreover, the spores of *P. echinosporum* are conspicuously spiny while those of *P. bivalve* are minutely warted. The fruiting bodies of *Physarum bogoriense* Racib., a somewhat similar species that is primarily tropical in distribution, have a peridium consisting of three layers and are not laterally compressed. *Physarum superbum* Hagelst. produces fruiting bodies that are laterally compressed to nearly terete, but this species can be easily distinguished by its bright ochraceous yellow to orange or orange-brown color. *Physarum bitectum* G. Lister [page 165, Figure 8-56] closely resembles *P. bivalve*, but the fruiting bodies are rounded or only slightly compressed. In *Physarum bitectum*, the spores are coarsely warted, while in *P. bivalve* the spores are minutely warted. It was recently determined that the original name proposed for this species and thus the correct one to use is *Angioridium sinuosum* (Bull.) Grev.

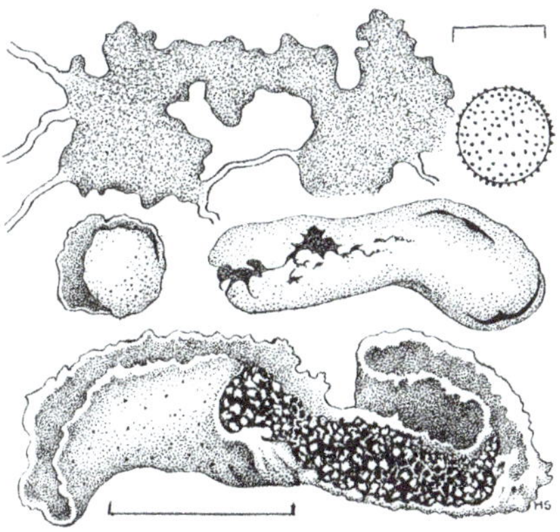

Figure 7-46B. *Physarum bivalve*. Three fruiting bodies of different sizes and shapes (lower portion of drawing). A single spore and a portion of the capillitium (upper portion of drawing).

Figure 7-47A. *Physarum cinereum* (©Yuri Novozhilov)

Physarum cinereum (Batsch) Pers.

DESCRIPTION: **Sporangia** sessile (often merging into short plasmodiocarps), closely gregarious to crowded or heaped, subglobose to pulvinate, 0.3 to 0.5 mm in diameter. **Hypothallus** inconspicuous. **Peridium** single, white to cinereous (iridescent to dark brown when limeless) membranous, more or less densely covered or impregnated with lime. **Capillitium** physaroid, consisting of numerous, variously shaped, mostly angular or branching white lime nodes (these occasionally massed in the center) connected by hyaline filaments. **Spores** purplish brown in mass, light violaceous by transmitted light, asperulate to minutely verrucose, 9 to 11 µm in diameter. **Plasmodium** watery-white or colorless.

SUBSTRATE: Dead leaves; also fruiting on living plants.

COMMENTS: Several other species of *Physarum* produce sessile fruiting bodies and fruit in similar situations. *Physarum rubiginosum* Fr. & Palmquist has rust-colored, reddish brown, or scarlet sporangia and dull orange to rusty brown lime nodes. *Physarum lateritium* (Berk. & Ravenel) Morgan is very similar in appearance to *P. rubiginosum*, from which it differs in its less intense color, more delicate peridium, and smaller spores (9 to 12 µm in *P. rubiginosum* and 7 to 9 µm in *P. lateritium*). The fruiting bodies of *Willkommlangea reticulata* (Alb. & Schwein.) Kuntze [page 171, Figure 8-74] superficially resemble those produced by several of the more plasmodiocarpous species of *Physarum*. The peridium of *W. reticulata*, however, is typically marked with prominent red spots, and the capillitium consists of two distinctly different systems. One of these systems is nearly limeless while the other consists of flat, angular lime nodes aggregated to form plate-like partitions that cause the interior of a plasmodiocarp to appear to be segmented. *Physarum contextum* (Pers.) Pers. has pale yellow, crowded sporangia with a double peridium and large (mostly 11 to 13 µm in diameter) spores. The sporangia of *P. luteolum* Peck are bright chrome-yellow and the lime nodes in the capillitium are yellow. *Physarum serpula* Morgan is typically much more plasmodiocarpous than *P. cinereum* and often forms a simple reticulum on the substrate. It is usually dull yellow or ochraceous and has a peridium consisting of only a single layer. *Physarum auriscalpium* Cooke appears to be closely related to *P. serpula* but differs in having the upper portion of the peridium dotted with glossy lime scales (as opposed to having a peridium evenly impregnated with lime as is the case in *P. serpula*). *Physarum auriscalpium* is infrequently encountered in nature but commonly appears on tree bark in moist chamber cultures. *Physarum decipiens* M.A. Curtis [Figure 8-59] is another species typically occurring on tree bark in moist chamber cultures. This species is similar to both *P. auriscalpium* and *P. serpula* but can be distinguished on the basis of its "bumpy-looking" peridium (due to the presence of prominent lime scales) and very large lime nodes in the capillitium.

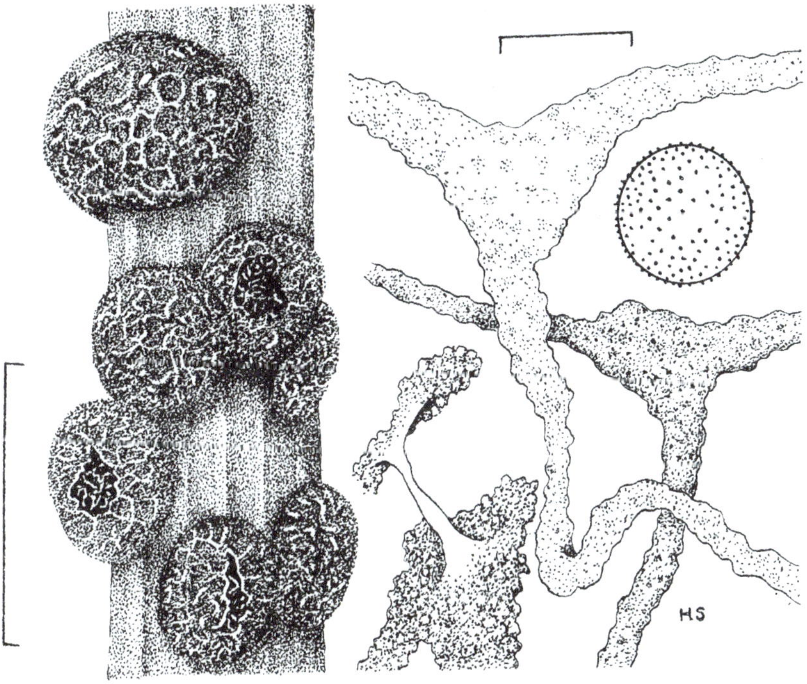

Figure 7-47B. *Physarum cinereum.* Several sporangia on a blade of grass (left side of drawing). A single spore and portions of the capillitium (right side of the drawing).

Figure 7-48A. *Physarum globuliferum* (©Joe Takano)

Physarum globuliferum (Bull.) Pers.

DESCRIPTION: **Sporangia** stalked, gregarious or sometimes united in small clusters, 0.6 to 1.5 mm tall. Sporotheca erect, globose, 0.4 to 0.7 mm. **Stalk** moderately slender, smooth or wrinkled, calcareous, usually white but sometimes ochraceous or dull red, 0.6 to 1.0 mm long. **Hypothallus** inconspicuous. **Peridium** white, pinkish white, gray, pale to dingy yellow, or ochraceous, single, membranous, densely encrusted with lime. **Columella** short, conical or blunt (rarely lacking). **Capillitium** physaroid, consisting of a dense network of small, rounded white or yellowish lime nodes connected by hyaline threads. **Spores** dark brown in mass, light violaceous brown by transmitted light, asperulate to verruculose, 7 to 9 μm in diameter. **Plasmodium** white or yellow.

SUBSTRATE: Decaying wood.

COMMENTS: This species can usually be identified by the combination of the calcareous stalk, the conical columella, and the dense capillitium that retains its shape even after the spores have been dispersed. *Physarum murinum* Lister, which is much less common, is pale brown in color and has brown lime nodes in the capillitium. *Physarum melleum* (Berk. & Broome) Massee [page 167, Figure 8-62] has yellow to dull orange sporangia with stout, typically white stalks and usually fruits on dead leaves. *Physarum galbeum* Wingate is somewhat smaller (usually more than 0.5 mm in diameter) than *P. globuliferum* and has golden yellow or chrome-yellow sporangia, slender noncalcareous stalks, and a capillitium in which the lime nodes are poorly differentiated. *Physarum galbeum* is not common in nature but sometimes appears on tree bark in moist chamber cultures. The sporangia of *P. tenerum* Rex are pale to bright yellow but have more or less calcareous stalks, a peridium that frequently dehisces into petal-like lobes, and a capillitium with well-differentiated lime nodes. *Physarum stellatum* (Massee) G.W. Martin also has sporangia with calcareous stalks and a distinctive pattern of dehiscence (the peridium splits into 6 to 12 petal-like lobes), but the sporangia are white and have a central lime mass (a pseudocolumella) in the sporotheca. A central lime mass is also present in *P. nucleatum* Rex, but this species has a noncalcareous, usually yellow to beige stalk and an irregular dehiscence. *Physarum pulcherripes* Peck resembles *P. globuliferum* but is red or yellow-orange to rusty brown in color. *Physarum penetrale* Rex is a distinctive species with ellipsoidal to prolate, often greenish yellow sporangia and long, delicate orange stalks. In this species, the top of the stalk extends upward into the spore mass as a rod-like, noncalcareous columella. *Physarum roseum* Berk. & Broome [page 168, Figure 8-64] is an even more distinctive species because of its bright red color. Interestingly, this species also has a plasmodium that is bright red. Based on recent molecular data, *Physarum globuliferum* was placed in the new genus *Nannengaella* as *N. globulifera* (Bull.) J.M. García-Martín, J.C. Zamora & Lado.

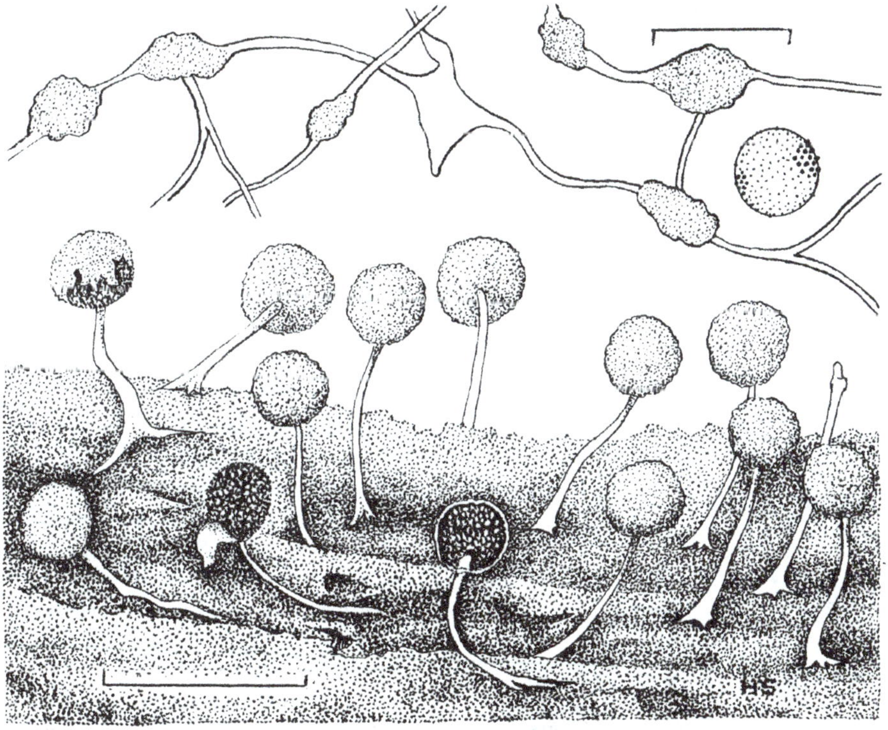

Figure 7-48B. *Physarum globuliferum*. Two groups of sporangia (lower portion of drawing). Note that one sporangium has been sectioned to show the conic columella. A single spore and portions of the capillitium (upper portion of drawing).

Figure 7-49A. *Physarum polycephalum* (©Mark Bower)

Physarum polycephalum Schwein.

DESCRIPTION: **Sporangia** stalked, gregarious, 1.5 to 2.0 mm tall. **Sporotheca** irregularly lenticular to discoid or convoluted to gyrose-confluent, 1.0 to 1.5 mm in diameter. **Stalk** slender, flexuous, twisted, sometimes flattened and expanded at the base, yellow, translucent, 1.0 to 2.0 mm long. **Hypothallus** membranous, inconspicuous. **Peridium** yellow to gray (rarely white), single, membranous, fragile, white or very pale yellow, marked with scattered clusters of yellow to white lime granules. **Capillitium** physaroid, dense, delicate, often expanding, consisting of a network of slender, hyaline filaments with membranous expansions at the junctions and varying quantities of yellow to white, fusiform to irregular lime nodes. **Spores** blackish brown in mass, violaceous by transmitted light, minutely and densely asperulate to spinulose, 8 to 11 µm in diameter. **Plasmodium** yellow or greenish yellow.

SUBSTRATE: Decaying wood and the fruiting bodies of fleshy fungi; sometimes fruiting on living plants.

COMMENTS: *Physarum polycephalum* has been widely used in laboratory studies and often has been used to exemplify the myxomycetes in general botany or biology textbooks. The species does not appear to be particularly common in nature, however. The sporangia of *P. compressum* Alb. & Schwein. [page 165, Figure 8-57] are occasionally somewhat contorted, but the usual expression is for the sporangia to be compressed-globose to more or less fan-shaped. Moreover, *P. compressum* has a relatively short, stout stalk that is either dark brown to almost black or coated with white lime (as opposed to translucent and yellow in *P. polycephalum*). *Physarum pezizoideum* (Jungh.) Pavill. & Lagarde produces sporangia in which the sporotheca is flat and saucer-shaped. Based on recent molecular data, *Physarum polycephalum* was placed in the genus *Badhamia* as *Badhamia polycephala* (Schwein.) J.M. Garcia-Martin, J.C. Zamora and Lado.

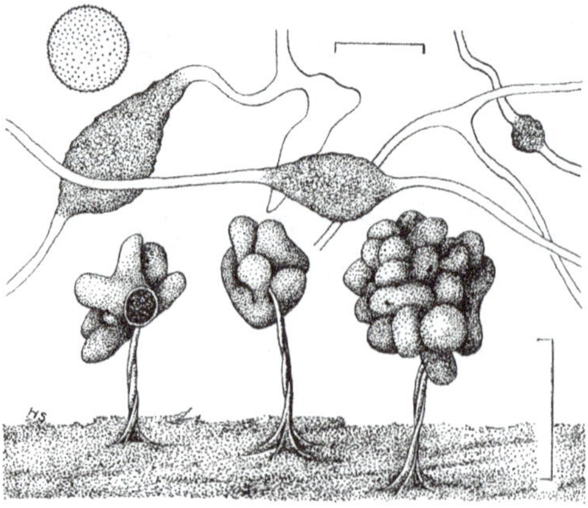

Figure 7-49B. *Physarum polycephalum*. Three sporangia (lower portion of drawing). A single spore and a portion of the capillitium (upper portion of drawing).

Figure 7-50A. *Physarum virescens* (©Raymond McNeil)

Physarum virescens Ditmar

DESCRIPTION: **Sporangia** sessile (or rarely stalked), crowded or superimposed in small clusters, spherical, ovate, or elongated, 0.2 to 0.4 mm in diameter. **Stalk** when present weak and strand-like, very short. **Hypothallus** colorless or pale brown, common to a group of sporangia. **Peridium** yellow, greenish yellow, or pallid gray-green, consisting of a single layer, thin, fragile, rugose, more or less densely calcareous. **Capillitium** physaroid, consisting of a delicate network of small, irregular, yellow lime nodes connected by hyaline filaments. **Spores** dingy brownish black in mass, pale violaceous gray by transmitted light, minutely punctate, 8 to 10 µm in diameter. **Plasmodium** bright yellow or yellow-green.

SUBSTRATE: Dead leaves; sometimes occurring on living plants.

COMMENTS: The distinctive characteristics of this species are the small, heaped, irregularly shaped, yellow or greenish yellow sporangia and the small, yellow lime nodes of the capillitium. The sporangia of *Physarum didermoides* (Pers.) Rostaf. [page 166, Figure 8-60] are white to bluish or purplish gray and have a peridium consisting of two layers. The outer layer is crustose, limy, and very fragile. This species also has white lime nodes and large (12 to 15 µm in diameter), densely spiny and somewhat angular spores. *Physarum didermoides* tends to fruit on the bark of decaying logs that are at a relatively early stage of decomposition. *Physarum contextum* (Pers.) Pers. is another species with densely massed sporangia. Typically pale yellow, the sporangia of *P. contextum* have a peridium consisting of two layers and spores that are 11 to 13 µm in diameter.

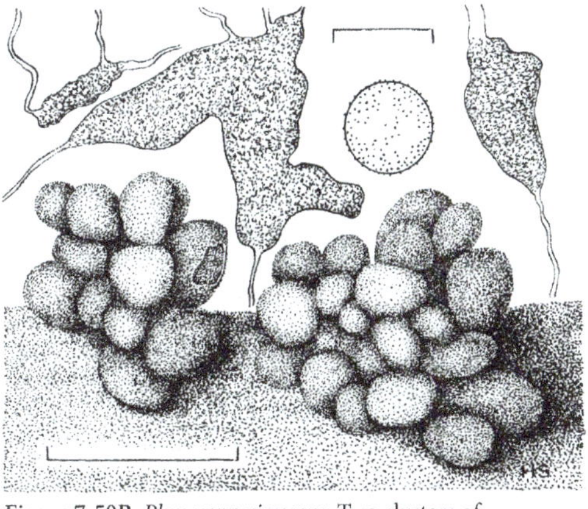

Figure 7-50B. *Physarum virescens*. Two clusters of sporangia (lower portion of drawing). A single spore and portions of the capillitium (upper portion of drawings).

Figure 7-51A. *Prototrichia metallica* (©John Shadwick)

Prototrichia metallica (Berkeley) Massee

DESCRIPTION: **Sporangia** sessile (or rarely short-stalked), scattered to gregarious, globose to subglobose, 0.5 to 2.2 mm in diameter. **Stalk** when present stout and dark, 0.1 to 0.3 mm long. **Hypothallus** inconspicuous. **Peridium** orange-brown to dull brown or sometimes rosaceous, thin, transparent, often somewhat iridescent. **Capillitium** consisting of numerous yellow-brown, more or less spirally banded threads that originate at the base of the sporangium and ultimately become somewhat subdivided above. **Spores** orange-brown to dull brown in mass, yellow by transmitted light, spiny, 10 to 14 µm in diameter. **Plasmodium** white.

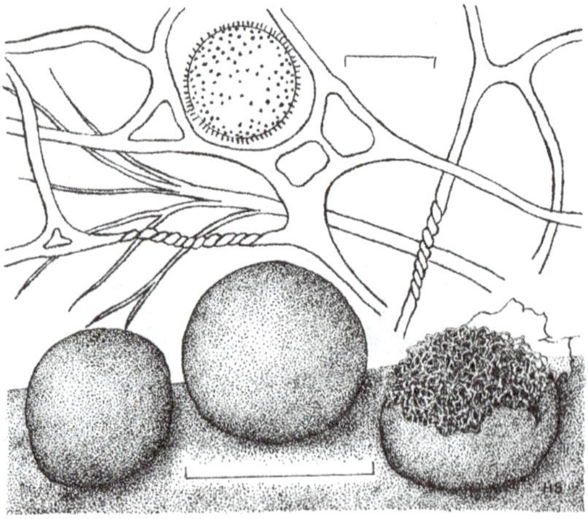

Figure 7-51B. *Prototrichia metallica*. Three sporangia (lower portion of drawing). A single spore and portions of the capillitium (upper portion of drawing).

SUBSTRATE: Decaying wood and bark, especially that of coniferous trees.

COMMENTS: *Prototrichia metallica* is one of the more distinctive myxomycetes characteristically found in alpine regions of the temperate zone. *Calomyxa metallica* (Berk.) Nieuwl. [page 151, Figure 8-14], a somewhat similar species that often appears on bark in moist chamber cultures, has much smaller sporangia and a capillitium consisting of very slender threads that are usually rather flexuous or coiled.

Figure 7-52A. *Reticularia splendens* (©Richard Orr)

Reticularia splendens Morgan

DESCRIPTION: **Aethalia** pulvinate to hemispherical, solitary to gregarious (although often somewhat separated), 0.5 to 6.0 cm broad and up to 1 cm thick. **Hypothallus** white, usually forming a conspicuous ring around the base of the aethalium. **Cortex** more or less tough-cartilaginous, reddish brown or umber, lustrous, smooth to rugose. **Pseudocapillitium** consisting of a system of anastomosing perforated plates that are sometimes frayed at the margins. **Spores** rust-brown in mass, pale brownish yellow or reddish brown by transmitted light, distinctly reticulate over about two-thirds of the surface and verruculose elsewhere, 7 to 9 µm in diameter. **Plasmodium** at first watery white and then changing through rose to brown.

SUBSTRATE: Decaying wood.

COMMENTS: This species is listed as *Enteridium splendens* (Morgan) T. Macbr. in some texts. Another commonly encountered species of *Reticularia* is *R. lycoperdon* Bull. [page 168, Figure 8-66]. The latter is usually larger than *R. splendens* and has a smooth, often silvery cortex. The morphologically similar *Licaethalium olivaceum* (Ehrenb.) Rostaf., formerly listed as *Reticularia olivacea* (Ehrenb.) Fr., is typically much smaller, olive-green to olive-brown in color, and has spores that occur in loose to compact clusters of 2 to 20.

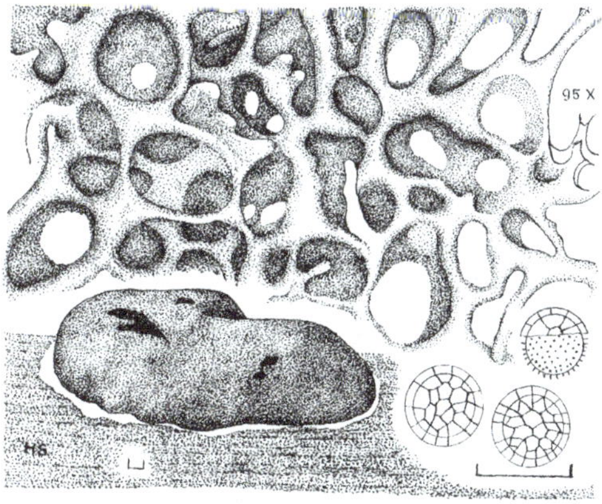

Figure 7-52B. *Reticularia splendens*. A single aethalium (lower left). Psuedocapillitium (upper portion of drawing). Several spores showing the variation that can exist (lower right).

Several other species of myxomycetes produce large aethalia but differ from *Reticularia splendens* in a number of important respects. *Brefeldia maxima* (Fr.) Rostaf. [page 151, Figure 8-13], reputed to produce the largest fruiting body (occasionally with a maximum extent of more than 50 cm across) of any myxomycete, has black spores and a capillitium that contains distinctive multicellular vesicles [page 73, Figure 6-61]. *Amaurochaete atra* (Alb. & Schwein.) Rostaf. [page 147, Figure 8-2] also has black spores, but there is no capillitium present. Instead, *A. atra* has a psuedocapillitium that consists of frayed, fan-like plates. Both *B. maxima* and *A. atra* appear to be uncommon, although the fact that the former tends to fruit late in the season, at least in North America, may account for why it is so rarely encountered.

Figure 7-53A. *Stemonitis axifera* (©Kim Fleming)

Stemonitis axifera (Bull.) T. Macbr.

DESCRIPTION: **Sporangia** stalked, fasciculate in small- or medium-sized clusters, 7 to 15 mm tall. **Sporotheca** erect, cylindrical, acuminate. **Stalk** black, shining, 3 to 7 mm long. **Peridium** not persisting in mature fruiting bodies, spore mass bright rusty brown but becoming pale brown as the spores are dispersed. **Hypothallus** membranous, inconspicuous. **Columella** branching freely and evenly, dissipated below the apex of the sporangium, the ultimate branchlets united into a delicate, small-meshed surface net. **Spores** bright reddish brown in mass, very pale reddish brown to almost colorless by transmitted light, nearly smooth or minutely punctate, 5 to 7 µm in diameter. **Plasmodium** white to pale or bright yellow.

SUBSTRATE: Decaying wood.

COMMENTS: This is a very common species of *Stemonitis* and usually can be distinguished on the basis of the bright rusty brown color alone. *Stemonitis smithii* T. Macbr., a closely related species, has smaller sporangia (usually no more than 6 mm tall) and smaller spores (4 to 5 µm in diameter). It should be noted that in some texts, *S. smithii* is not considered as distinct from *S. axifera*. The sporangia of *S. flavogenita* E. Jahn are darker in color and tend to have conspicuous membranous expansions in the capillitium as well as larger (7 to 9 µm in diameter) and somewhat darker, verruculose spores. Moreover, this species is more likely to fruit on leaf litter and other plant debris on the forest floor than on decaying wood, the usual substrate for *S. axifera*. *Stemonitis herbatica* Peck also typically fruits on leaf litter and various types of plant debris as well as on living plants. *Stemonitis herbatica* and *S. flavogenita* are very similar species, and sometimes apparently intermediate forms are collected. *Stemonitis flavogenita*, however, usually can be distinguished by its longer stalks and the presence of the distinctive membranous expansions in the capillitium. *Stemonitis virginiensis* Rex, a relatively uncommon species, is similar in appearance to *S. axifera* but is readily distinguished by its sharply banded-reticulate spores.

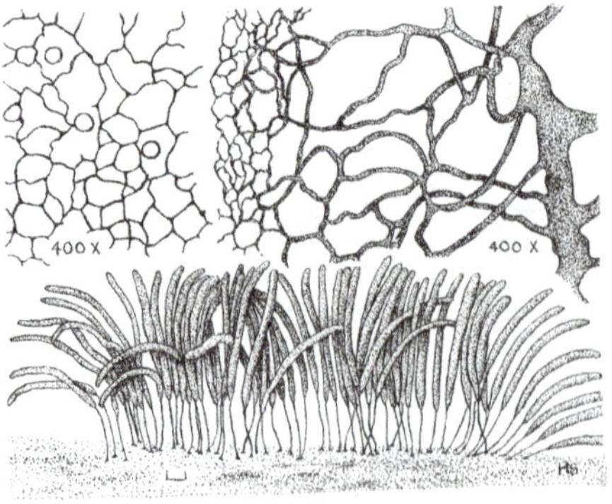

Figure 7-53B. *Stemonitis axifera*. A cluster of sporangia (lower portion of drawing). A semidiagrammatic sketch showing the relationship of the capillitium to the columella and surface net (upper right). Several spores and a portion of the surface net (upper right).

Figure 7-54A. *Stemonitis fusca* (©Todd Elliott)

Stemonitis fusca Roth

DESCRIPTION: **Sporangia** stalked, tufted, often in large colonies, 6 to 20 mm tall. **Sporotheca** slender, cylindrical. **Stalk** black, shining, rather long, from nearly one-half to one-fourth or less of the total height. **Peridium** not persisting in mature fruiting bodies, spore mass deep fuscous to dark reddish brown, becoming paler as the spores are dispersed. **Hypothallus** membranous, shiny brown, common to a group of sporangia. **Peridium** not persisting in mature sporangia. **Columella** dark brown or black, reaching the apex of the sporangium. **Capillitium** arising from all parts of the columella, branching and anastomosing freely, the ultimate branchlets united into a close-meshed surface net. **Spores** fuscous in mass, violet-brown by transmitted light, prominently to delicately warted-reticulate, 7.5 to 9.0 µm in diameter. **Plasmodium** white.

SUBSTRATE: Decaying wood.

COMMENTS: This is a common and usually rather distinctive species. *Stemonitis splendens* Rostaf. [page 169, Figure 8-68] often occurs in larger fruitings and has sporangia with a conspicuously large-meshed surface net and nonreticulate spores. The sporangia of *S. nigrescens* Rex are appreciably smaller (2 to 5 mm tall) and have shorter stalks than those of *S. fusca*. However, the former is not always considered as distinct from the latter. *Symphytocarpus trechisporus* (Berk. ex Torrend) Nann.-Bremek. (listed as *Stemonitis trechispora* [Berk. ex Torrend]) T. Macbr. in earlier treatments of the myxomycetes) has relatively small (3 to 7 mm tall), sessile or very short-stalked, very dark sporangia. The large

(11 to 13 μm in diameter), prominently reticulate spores represent the most distinguishing feature of this relatively uncommon species. Fruitings of *Stemonaria longa* (Peck) Nann.-Bremek., R. Sharma & Y. Yamam. (listed as *Comatricha longa* Peck in earlier treatments of the myxomycetes) [page 169, Figure 8-67], a species more common in the tropics than in temperate regions, might be mistaken for *S. fusca* in the field; however, the sporangia of the former tend to be drooping or pendant, the spores are darker, and the capillitium is entirely different. The primary branches that arise from the columella of *S. longa* are rigid, mostly free from one another, and dichotomously forked.

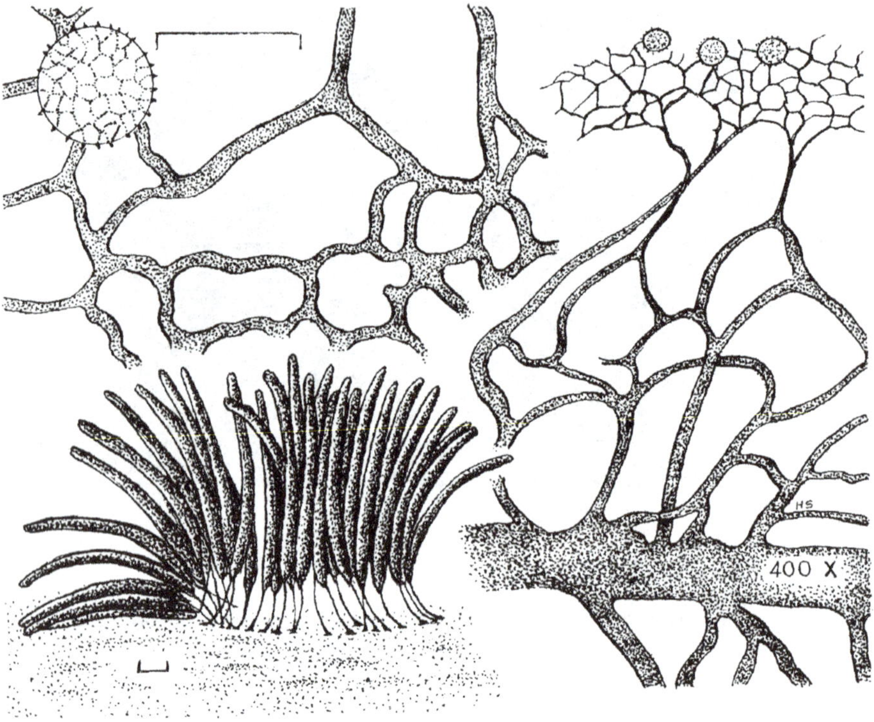

Figure 7-54B. *Stemonitis fusca.* A cluster of sporangia (lower left). A semidiagrammatic sketch showing the relationship of the capillitium to the columella and surface net (right side of drawing). A single spore and a portion of the surface net (upper left).

Figure 7-55A. *Stemonitopsis typhina* (©Juan Francisco Moreno)

Stemonitopsis typhina (H.G. Wigg.) Nann.-Bremek.

DESCRIPTION: **Sporangia** stalked, scattered to gregarious, 2 to 5 mm tall. **Sporotheca** erect cylindrical to somewhat ovate and tapering above, 0.2 to 0.6 mm in diameter. **Stalk** slender, black, typically about half the height of the sporotheca, often covered with a silvery film. **Hypothallus** reddish brown, contiguous for a group of sporangia. **Peridium** membranous, silvery lilac-gray, usually persisting in mature sporangia as scattered patches or sometimes as a shallow calyculus at the base of the spore mass. **Capillitium** a dense network of pale brown threads with numerous anastomoses. **Spores** brown in mass, very pale lilac-brown by transmitted light, faintly punctate with scattered clusters of dark warts, 6 to 8 µm in diameter. **Plasmodium** translucent white.

SUBSTRATE: Decaying wood and (less commonly) bark.

COMMENTS: The clusters of dark warts on the spores are usually quite evident and make this a relatively easy species to identify. *Stemonitopsis aequalis* (Peck) Y. Yamam. is similar but usually occurs in clusters in which the individual sporangia are more evenly spaced, the spore mass is darker than that of *S. typhina*, and the spores lack the clusters of dark warts. *Stemonitopsis aequalis* is most likely to be encountered in high-elevation coniferous forests. The sporangia of the rather common *Comatricha pulchella* (C. Bab.) Rostaf. [page 153, Figure 8–19] are pale brown and typically occur on dead leaves and other types of plant debris rather than on decaying wood, which is the usual substrate for both *S. aequalis* and *S. typhina*. The sporangia of *C. laxa* Rostaf., a species that can be found on wood or litter and also occurs on bark in moist chamber cultures, are typically ovate to short-cylindrical and reddish brown. Moreover, the capillitium of *C. laxa* is rather open and the columella gives rise to branches that are more or less horizontal. *Comatricha suksdorfii* Ellis & Everh. [page 153, Figure 8-20], a species characteristic of alpine areas, has large (4 to 8 mm tall), robust sporangia and dark, prominently warted spores that are mostly 10 to 12 µm in diameter. *Symphytocarpus confluens* (Cooke & Ellis) Ing & Nann.-Bremek. occurs in similar situations but can be recognized by its shorter (only 2 to 4 mm tall), much more densely aggregated sporangia. *Stemonitopsis hyperopta* (Meyl.) Nann.-Bremek., which seems to be most abundant in high-elevation coniferous forests, has lilaceous brown sporangia and small (5 to 6 µm in diameter), delicately reticulate spores.

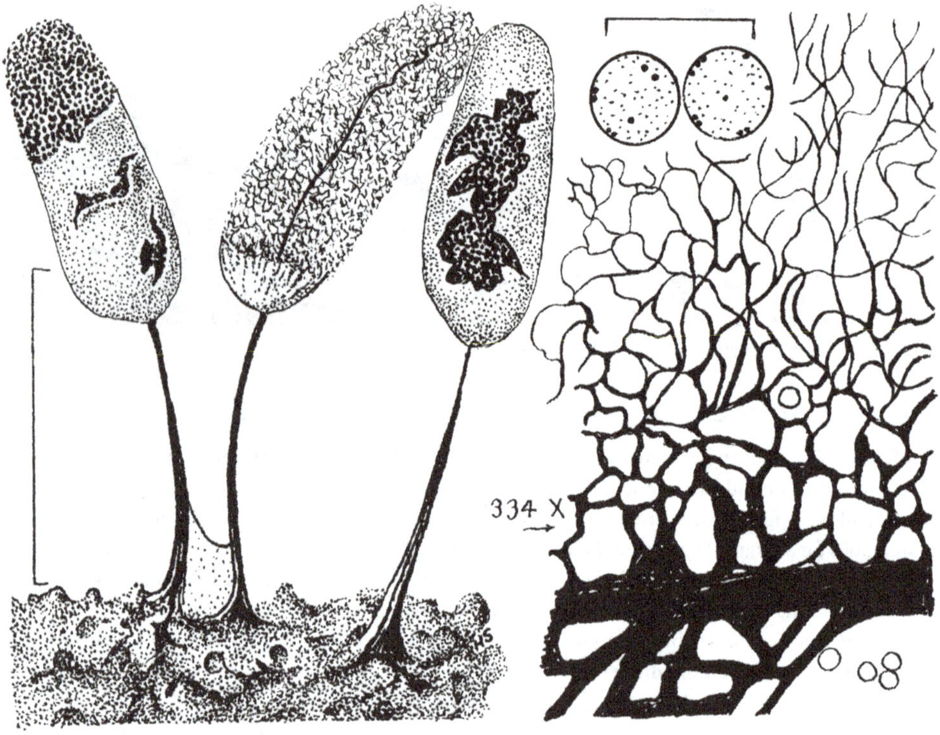

Figure 7-55B. *Stemonitopsis typhina*. Three sporangia showing the different stages of dehiscence (left slide of drawing). A semidiagrammatic sketch showing the relationship of the capillitium and columella (left side of drawing). Two spores showing the characteristic clusters of warts (upper right).

Figure 7-56A. *Trichia decipiens* (©Alain Michaud)

Trichia decipiens (Pers.) T. Macbr.

DESCRIPTION: **Sporangia** stalked (or rarely sessile), gregarious to crowded, up to 3 mm tall. **Sporotheca** turbinate to pyriform, 0.6 to 0.8 mm in diameter. **Stalk** cylindrical, furrowed, dark brown below and paler above, filled with spore-like cells, up to 1.0 mm long. **Hypothallus** inconspicuous. **Peridium** firm or membranous, dull yellow to olivaceous yellow or brown, often translucent when thin, persisting below as a deep (or sometimes rather shallow) calyculus. **Capillitium** consisting of free elaters, these olivaceous yellow, simple or branched, bearing three to five spiral bands, smooth, 5 to 6 µm in diameter, tapering gradually to the long slender tips. **Spores** olivaceous yellow in mass, pale yellow by transmitted light, delicately reticulate, 10 to 13 µm in diameter. **Plasmodium** white or pink.

SUBSTRATE: Decaying wood, especially that of coniferous trees.

COMMENTS: Three other species of *Trichia* with stalked sporangia are *T. botrytis* (J.F. Gmel.) Pers. [page 167, Figure 8-69], *T. erecta* Rex, and *T. verrucosa* Berk. [page 171, Figure 8-73]. The first two species produce sporangia that typically are somewhat clustered, but they usually can be distinguished on the basis of spore color. *Trichia botrytis* has dull yellow to dingy ochraceous spores, while those of *T. erecta* are bright yellow. *Trichia verrucosa* also has bright yellow spores, but the spore ornamentation of the two species is quite different (prominently reticulate in *T. verrucosa* versus minutely warted in *T. erecta*). Another difference that exists among these three species is that the elaters of *T. erecta* and *T. verrucosa* have short-tapering ends, while those of *T. botrytis* have long-tapering ends. *Trichia subfusca* Rex [page 170, Figure 8-72], another species characterized by elaters with short-tapering ends, has bright straw-yellow spores and a short, stout, dark brown or black stalk. Sporangia of *Hemitrichia pardina* (Minakata) Ing [page 107] in which the stalk is well developed somewhat resemble the sporangia of several stalked species of *Trichia*. However, *H. pardina* can be distinguished by the conspicuous dark, peg-like warts that occur on the upper portion of the peridium. Based on recent molecular data, *T. decipiens* was placed in the genus *Hemitrichia* as *H. decipiens* (Pers.) García-Cunch., J.C. Zamora & Lado.

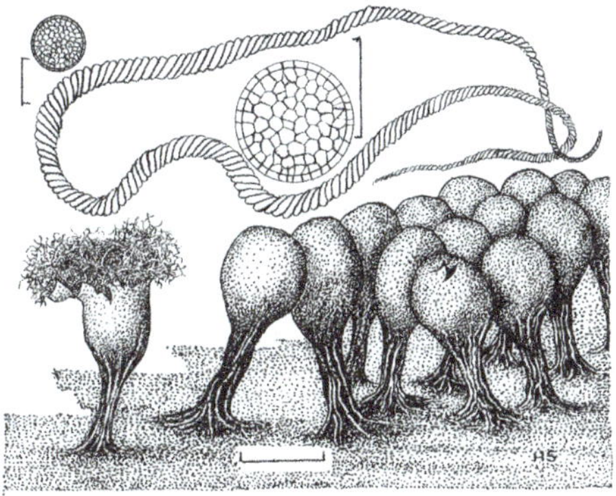

Figure 7-56B. *Trichia decipiens*. Group of sporangia (lower portion of drawing). Two spores and isolated capillitial strand (upper portion of drawing). Note that the spores are show at two different magnifications while the elater is shown at the lower magnification only.

Figure 7-57A. *Trichia favoginea* (©Kim Fleming)

Trichia favoginea (Batsch) Per.

DESCRIPTION: **Sporangia** sessile, usually densely crowded but occasionally gregarious, subglobose to ovate or cylindrical, 0.5 to 1.0 mm in diameter and up to 2 mm tall. **Hypothallus** membraneous, colorless. **Peridium** bright yellow-brown, more or less shining membranous, transparent, yellow, nearly smooth. **Capillitium** yellow, consisting of elaters 4 to 8 µm in diameter, marked with three to five smooth to spinulose spiral bands, tips of elaters pointed, blunt, or rounded and spinulose. **Spores** yellow to ochraceous or orange in mass, pale to bright yellow by transmitted light, coarsely reticulate, 13 to 15 µm in diameter. **Plasmodium** white.

SUBSTRATE: Decaying wood or bark; occasionally found on dead leaves.

COMMENTS: The coarsely reticulate spores distinguish *Trichia favoginea* from all other sessile species of *Trichia*. *Oligonema flavidum* (Peck) Peck [page 163, Figure 8-51] and *O. schweinitzii* (Berk.) G.W. Martin also have coarsely reticulate spores and, because of their sessile, yellow sporangia, superficially resemble *T. favoginea*. There is a tendency, however, for the sporangia of both species of *Oligonema* to be densely crowded or heaped, while the sporangia of *T. favoginea* generally form a single compact layer. Moreover, the capillitium, which is usually much less abundant in the two species of *Oligonema*, consists of generally shorter elaters that are nearly smooth or marked with faint spiral bands. The two species of *Oligonema* are best distinquished on the basis of spore characteristics; the surface reticulum on the spores of *O. flavidum* consists of narrow bands, while that of *O. schweinitzii* consists of rather broad bands. *Trichia alpina* (R.E. Fr.) Meyl., T. contorta (Ditmar) Rostaf. [page 170, Figure 8-70], and *T. scabra* Rostaf. [page 170, Figure 8-71] are three other species of *Trichia* that produce sessile sporangia. *Trichia scabra* has delicately reticulate spores and sporangia that are typically dull orange in color. *Trichia alpina*, a snowbank species, has dark purple to black sporangia and rather large (14 to 18 µm in diameter) spores and capillitium (6 to 8 µm in diameter). The sporangia of *T. contorta* are occasionally as dark as those of *T. alpina*, but the former has smaller (10 to 13 µm in diameter) spores and capillitium (3 to 5 µm in diameter).

The concept of *Trichia favoginea* used herein follows that of Farr (1958), who did not distinguish this species from the very similar *T. affinis* de Bary and *T. persimilis* P. Karst. All three are considered distinct species in some monographs (e.g., Stephenson 2021). On the basis of recent molecular data, *T. favoginea* was placed in the genus *Oligonema* as *O. favogineum* (Batsch) García-Cunch., J.C. Zamora & Lado.

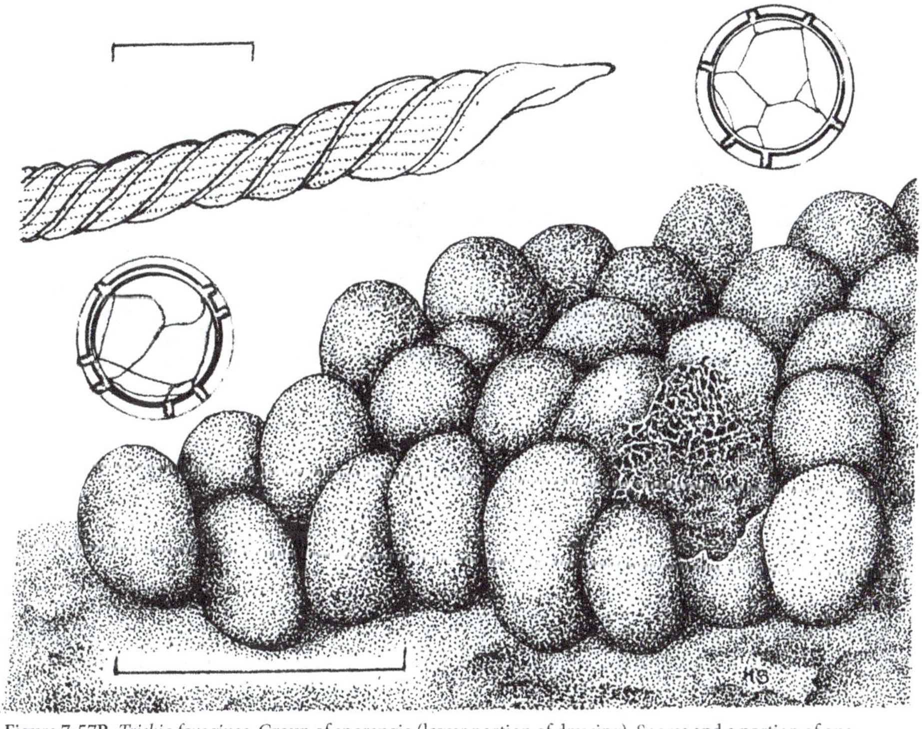

Figure 7-57B. *Trichia favoginea*. Group of sporangia (lower portion of drawing). Spores and a portion of one capillitial strand (upper portion of drawing).

Figure 7-58A. *Trichia varia* (©Alain Michard).

Trichia varia (Per. ex J.F. Gmel.) Pers.

DESCRIPTION: **Sporangia** sessile (or occasionally short-stalked) or short plasmodiocarps, gregarious to crowded, globose or obovate to somewhat elongated, rarely somewhat top-shaped when stalked, 0.5 to 0.9 mm in diameter and up to 1.0 mm tall. **Stalk** when present short, black, 0.1 to 0.4 mm long. **Hypothallus** contiguous for groups of sporangia but mostly inconspicuous. **Peridium** yellow to olive-brown to yellow-brown, persistent, membranous, shining or sometimes encrusted with darker material. **Capillitium** yellow to ochraceous, consisting of relatively long, free elaters, 3 to 5 µm in diameter, bearing two (or rarely three) prominent widely spaced spiral bands, tips acute, curved, about twice the diameter of the elater in length. **Spores** yellow to orange-yellow in mass, dull yellow by transmitted light, delicately warted, 12 to 14 µm in diameter. **Plasmodium** white.

SUBSTRATE: Decaying wood; less commonly on bark.

COMMENTS: *Trichia varia* is both common and easy to recognize from the widely spaced spiral bands on the capillitium. Like many species of *Trichia*, fruitings of this species tend to occur late in the season. *Trichia contorta* (Ditmar) Rostaf. [page 170, Figure 8-70] is morphologically similar to *T. varia* in that it produces sporangia and short plasmodiocarps. However, the capillitium of *T. contorta* has 4 to 5 spiral bands. Several varieties of *T. contorta* have been recognized, based on minor differences in the capillitium. The sporangia of *Colloderma oculatum* (C. Lippert) G. Lister [page 151, Figure 8-18] and *Colloderma robustum* (G. Lister ex Meyl.) Meyl. are similar in size and shape to those of *T. varia*, but the two species differ in virtually every other respect. For example, in *T. varia* the spores are yellow, while those of the two species of *Colloderma* are dark brown to black. All three species can be found on decaying wood, although this is not the usual substrate for either *C. oculatum* or *C. robustum*. The latter are more likely to be associated with mosses on wet rocks or moist decaying coniferous logs with a cover of leafy liverworts present. The distinguishing feature of both species of *Colloderma* is the thick gelatinous layer of the peridium. Before the latter dries out, the entire structure has the general appearance of an "eye" peeping out of the substrate upon which the sporangium occurs.

Figure 7-58B. *Trichia varia*. Several sporangia (lower portion of drawing). A single spore and an isolated capillitial strand (upper portion of drawing).

Figure 7-59A. *Tubifera ferruginosa* (©Kim Fleming)

Tubifera ferruginosa (Batsch) J.F. Gmel.

DESCRIPTION: **Pseudoaethalia** composed of numerous sessile sporangia crowded together, individual sporangia cylindrical to ovate, up to 0.4 mm in diameter and 5 mm tall, the entire structure reaching 15 cm or more in extent. **Hypothallus** well developed, membranous to spongy, colorless or pallid. **Peridium** membranous, thin, pale umber to reddish brown or purplish brown, persisting in mature fruiting bodies except at the apices of the individual sporangia, where it tends to break away. **Hypothallus** well developed, membranous to spongy, colorless or pallid. **Spores** umber-brown in mass, pallid by transmitted light, finely reticulate over about three-fourths of the surface, 6 to 8 µm in diameter. **Plasmodium** orange, apricot or pink but sometimes starting out as white.

SUBSTRATE: Decaying wood or wood debris; occasionally occurring on forest floor leaf litter.

COMMENTS: *Tubifera microsperma* (Berk. & M.A. Curtis) G.W. Martin and *Siphoptychium casparyi* Rostaf. (listed as *T. casparyi* [Rostaf.] T. Macbr. in earlier treatments of the myxomycetes) are the two species most likely to be confused with *T. ferruginosa.* The former can be distinguished by its stalk-like hypothallus, generally smaller size, and by its smaller (4.5 to 5.5 µm in diameter) spores. In *S. casparyi* a pseudocapillitium (which takes the form of a columella-like central body) is present in many of the individual sporangia. *Alwisia bombarda* Berk. & Broome (listed as *Tubifera bombarda* [Berk. & Broome] G.W. Martin [page 147, Figure 8-1] in earlier treatments of the myxomycetes), a striking species known only from the tropics, is stalked and has a pseudocapillitium consisting of a cluster of rigid, bristle-like filaments arising from the center of each sporangium. Although *A. bombarda* appears to be restricted to the tropics, other species in the genus have been described from material collected in temperate regions of Australia (Leontyev et al. 2014b).

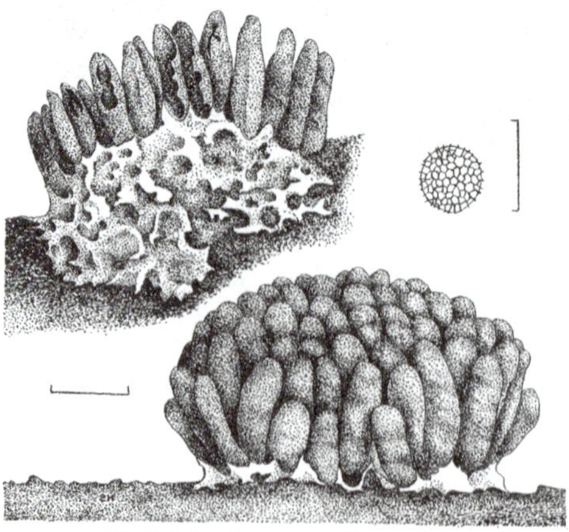

Figure 7-59B. *Tubifera ferruginosa*. External view of a pseudoaethalium (lower portion of drawing). Vertical section of same (upper portion of drawing). A single spore (upper right).

CHAPTER 8

Additional Species

This chapter contains an image and a brief description of 74 other species of myxomycetes likely to be encountered in the field or in moist chamber cultures.

Alwisia bombarda Berk. & Broome (Figure 8-1)

Figure 8-1. *Alwisia bombarda* (©Carlos Rojas)

In most treatments of the myxomycetes, this species is listed as *Tubifera bombarda* (Berk. & Broome) G.W. Martin although it was originally described as *Alwisia bombarda*. The stalked pseudoaethalium with the tufts of bristle-like structures that project outward from each sporangium-like unit is so distinctive that this species is unlikely to be confused with any other myxomycete. *Alwisia bombarda* occurs in tropical forests, where it is found on decaying wood.

Amaurochaete atra (Alb. & Schwein.) Rostaf. (Figure 8-2)

Figure 8-2. *Amaurochaete atra* (©Henri Koskinen)

The fruiting bodies (aethalia) of this species are quite large, sometimes reaching 8 cm in total extent. *Amaurochaete atra* is associated with the decaying wood of conifers, especially pine (*Pinus* spp.). At first the cortex is shining black, but it soon becomes dull black and exceedingly fragile, which allows the spore mass to be easily dispersed. As is the case for all myxomycetes that produce large fruiting bodies, those of *A. atra* are likely to attract slime mold beetles.

Arcyria globosa Schwein. (Figure 8-3)

Figure 8-3. *Arcyria globosa* (©Joe Takano)

This species almost invariably occurs on a very specific substrate—the spines of decaying burs of chestnut trees (*Castanea* spp.). The scattered white (when fresh) fruiting bodies with a deep calyculus are distinctive. Because the American chestnut (*C. dentata* [Marsh.] Broch.) native to eastern North America was almost completely eliminated by the chestnut blight fungus, *Arcyria globosa* has become rare. However, it occasionally occurs on the decaying burs of introduced species of *Castanea*.

Figure 8-4. *Arcyria incarnata* (©Alain Michaud)

Figure 8-5. *Arcyria obvelata* (©Henri Koskinen)

Figure 8-6. *Arcyria pomiformis* (©Henri Koskinen)

Arcyria incarnata (Pers. ex J.L. Gmel.) Pers. (Figure 8-4)

The fruiting bodies of this species are very similar to those of *Arcyria denudata* (page 77), but the capillitium is only loosely attached to the calyculus and is very easily detached. In *A. denudata* the capillitium is strongly attached to the calyculus. Both *A. incarnata* and *A. denudata* are among the "red" species of *Arcyria*, but the fruiting bodies of the latter weather to brown in older specimens. Like *A. denudata*, *A. incarnata* occurs on decaying wood.

Arcyria obvelata (Oeder) Onsberg (Figure 8-5)

This species is listed as *Arcyria nutans* (Bull.) Grew. in most earlier treatments of the myxomycetes. The short-stalked fruiting bodies are pale yellow when fresh and occur in dense clusters on decaying wood. The single most distinctive feature is the capillitium, which expands to become a drooping, plume-like structure. The capillitium readily detaches from the calyculus, which is deep and vase-like.

Arcyria pomiformis (Leers) Rostaf. (Figure 8-6)

Like that of *Arcyria globosa* (page 147, Figure 8-3), the sporotheca of *A. pomiformis* is globose, but when fresh it is ochraceous to yellow and never white. Moreover, the latter species occurs on decaying wood or sometimes appears on bark in moist chamber cultures and not on chestnut burs. The fruiting bodies of *A. pomiformis* tend to be scattered on the substrate upon which they occur.

Arcyria stipata (Schwein.) Lister (Figure 8-7)

This species can be recognized by the closely appressed to superimposed (almost "heaped") fruiting bodies. In fact, based on the general appearance of a group of fruiting bodies, one might consider it to be a pseudoaethalium. The short-stalked fruiting bodies have a copper-colored, somewhat metallic, and persistent peridium. *Arcyria stipata* occurs on decaying wood. It does not appear to be particularly common.

Figure 8-7. *Arcyria stipata* (©Henri Koskinen)

Badhamia affinis Rostaf.

(Figure 8-8)

This species typically produces small white, discoid, sessile fruiting bodies with a depressed center. Although usually sporangia, short plasmodiocarpous fruiting bodies also occur. *Badhamia affinis* is a species associated with the bark of living and dead trees. It is more likely to be encountered in a moist chamber culture than collected in the field. The large (13 to 15 μm) and spiny spores separate *B. affinis* from other morphologically similar species in the genus *Badhamia*

Figure 8-8. *Badhamia affinis* (©Alain Michaud)

Badhamia foliicola Lister

(Figure 8-9)

Relatively few myxomycetes commonly produce fruiting bodies on living plants, but it is not uncommon for *Badhamia foliicola* to be found on this substate. However, the species also occurs on various types of non-living plant debris. The fruiting bodies are small (usually no more than 0.5 mm in diameter), globose to somewhat elongated and then plasmodiocarpous, have a thin gray iridescent peridium, and often occur in extensive fruitings.

Figure 8-9. *Badhamia foliicola* (©Alain Michaud)

Figure 8-10. *Badhamia melanospora* (©Alain Michaud)

Figure 8-11. *Badhamia nitens* (©Jaun Francisco Moreno)

Figure 8-12. *Badhamia panicea* (©Jaun Francisco Moreno)

Badhamia melanospora Speg. (Figure 8-10)

This species is listed as *Badhamia gracilis* (T. Macbr.) T. Macbr. in most earlier treatments of the myxomycetes. The fruiting body consists of a globose to somewhat pear-shaped sporotheca on a thin, weak stalk. The sporotheca is umbilicate below and the peridium is thin and white. *Badhamia melanospora* is often abundant in deserts, where it occurs on old, decaying portions of cacti and other succulents.

Badhamia nitens Berk. (Figure 8-11)

The fruiting bodies of *Badhamia nitens* are sessile, globose to slightly flattened, and up to 1 mm in diameter. However, the most striking feature is the yellow color of the peridium, which has a "flaky" calcareous outer layer. Moreover, *B. nitens* is one of the species in the genus with tightly clustered spores (page 24, Figure 2-17). This is not a common species and is associated with the bark of living trees, often with mosses present. It also appears on bark from living trees places in moist chamber cultures.

Badhamia panicea (Fr. Rostaf. (Figure 8-12)

The fruiting bodies of *Badhamia panicea* are sessile, subglobose or slightly flattened sporangia or short plasmodiocarps, 0.4 to 2.0 mm in diameter, usually crowded, and typically occurring in large fruitings. Unlike some species of *Badhamia*, the spores are free. The peridium is white and covered with a thick layer of lime. This species is commonly found on the bark of fallen trees but does not appear in moist chamber cultures on the bark of living trees.

Brefeldia maxima (Fr.) Rostaf. (Figure 8-13)

The fruiting bodies (aethalia) of this species are the largest produced by any myxomycete, with some examples exceeding a meter in total extent. The cortex of athe fruiting body is purple-black and very fragile. The single most distinctive feature of this species other than the large size is the presence of multicellular vesicles (Figure 6-6) that occur at junctions of the capillitium. *Brefeldia maxima* is commonly found on decaying stumps and at the base of standing dead trees.

Figure 8-13. *Brefeldia maxima* (©Henri Koskinen)

Calomyxa metallica (Berk.) Nieuwl. (Figure 8-14)

The sessile fruiting bodies of *Calomyxa metallica* are usually more-or-less globose sporangia but plasmodiocarpous forms also occur. The distinguishing features are the thin membranous iridescent peridium and the yellow spores. Although *C. metallica* can be found on decaying wood, it is more likely to be encountered on bark of living trees placed in moist chamber cultures.

Figure 8-14. *Calomyxa metallica* (©Alain Michaud)

Ceratiomyxa morchella A.L. Welden (Figure 8-15)

Ceratiomyxa fruticulosa (page 81) has an almost worldwide distribution but two other species of *Ceratiomyxa* are restricted to the tropics. One of these is *C. morchella*, in which the fruiting bodies resemble miniature morels and are about 2 to 3 mm tall. These occur on decaying wood, usually that of large logs, in tropical forests.

Figure 8-15. *Ceratiomyxa morchella* (©Carlos Rojas)

Figure 8-16. *Ceratiomyxa sphaerosperma* (©Carlos Rojas)

Figure 8-17. *Collaria arcyrionema* (©Joe Takano)

Figure 8-18. *Colloderma oculatum* (©Joe Takano)

Ceratiomyxa sphaerosperma Boedijn (Figure 8-16)

Ceratiomyxa sphaerosperma, like *C. morchella* (page 151, Figure 8-15), is restricted to the tropics, where the fruiting bodies typically occur on decaying fruits with a hard outer covering but also can be found on other types of plant debris. The fruiting body consists of a gelatinous stalk bearing a cluster of branch-like structures at the apex. The entire fruiting body is only about 1 to 2 mm tall.

Collaria arcyrionema (Rostaf.) Nann.-Bremek.) ex Lado (Figure 8-17)

This species is listed as *Lamproderma arcyrionema* Rostaf. in most earlier treatments of the myxomycetes. The distinguishing features are the globose sporotheca and the relatively long (usually at least two-thirds of the total height) stalk. *Collaria arcyrionema* is similar in appearance to *Comatricha nigra* (page 84) but has a membranous silvery peridium that is lacking in the latter species. However, it should be noted that the peridium is often lacking in older specimens. *Collaria arcyrionema* occurs on decaying wood, often in large fruitings.

Colloderma oculatum (Lippert) G. Lister (Figure 8-18)

The slightly pulvinate or hemispherical fruiting body of *Colloderma oculatum* is immediately recognizable during the latter stages of its development but before it dries out. The peridium consists of a thick, gelatinous and somewhat transparent outer layer and a thin dark inner layer that confines the spore mass. This darker structure surrounded by the gelatinous layer, as viewed from above, has the appearance of an "eye" peering out of the substrate upon which it occurs. This species is found on moist decaying logs in montane coniferous forests and in association with bryophytes on the bark of living trees and on wet rocks.

Comatricha pulchella (C. Bab.) Rostaf. (Figure 8-19)

The distinguishing features of the fruiting bodies of *Comatricha pulchella* are the relatively short stalk (no more than half the total height), the ovoid sporotheca, the intricately branched capillitium, and the pinkish brown spores. The usual substrate for this species is leaf litter, but it sometimes occurs on decaying wood.

Comatricha suksdorfii Ellis & Everh. (Figure 8-20)

Comatricha suksdorfii is one of the more robust members of the genus *Comatricha*. The fruiting bodies are as much as 8 mm tall and usually occur in dense clusters on the decaying wood of conifers. The sporotheca in this species is cylindrical to ovate and the spore mass is black. This is an alpine species that is not uncommon in mountains such as those in the western United States.

Craterium aureum (Schumach.) Rostaf.

(Figure 8-21)

Craterium aureum, like other members of the genus *Craterium*, has a fruiting body with a more-or-less cup-shaped sporotheca after dehiscence. The bright yellow color of the peridium separates *C. aureum* from species such as *C. leucocephalum* (page 154, Figure 8-22) and *C. minutum* (page 86). The usual substrate for *C. aureum* is leaf litter.

Figure 8-19. *Comatricha pulchella* (©Juan Francisco Moreno)

Figure 8-20. *Comatricha suksdorfii* (©Alain Michaud)

Figure 8-21. *Craterium aureum* (©Alain Michaud)

Craterium leucocephalum (Pers.) Ditmar (Figure 8-22)

The fruiting body of *Craterium leucocephalum* is a stalked sporangium with a more-or-less cylindrical sporotheca, bicolored peridium (orange or brown below and white above), and a stalk that is about half the total height. At dehiscence, the top of the sporotheca is lost, leaving behind a deep cup with a ragged rim. This species in found on leaf litter and is often found in large fruitings.

Figure 8-22. *Craterium leucocephalum* (©Dmitro Leontyev)

Craterium obovatum Peck.

(Figure 8-23)

In some treatments of the myxomycetes this species is listed as *Badhamia obovata* (Peck) S.J. Sm. The distinguishing features are the sporangium with grayish to purplish peridium, and a columella that extends up to two-thirds the height of the sporotheca. This species can be found on decaying wood or various types of plant debris. On the basis of recent molecular data, *Craterium obovatum* was placed in the genus *Diachea as D. obovata* (Peck) J.M. García-Martín, J.C. Zamora & Lado.

Figure 8-23. *Craterium obovatum* (©Alain Michaud)

Cribraria confusa Nann.-Bremk & Y. Yamam.

(Figure 8-24)

Cribraria confusa is one of the smaller species of *Cribraria* and can be easily recognized from its golden yellow color and the absence of distinct nodes in the peridial net. This species develops on bark from living trees placed in moist chamber cultures. It is not uncommon to observe the dried spores as a round mass in the center of the space enclosed by the peridial net.

Figure 8-24. *Cribraria confusa* (©Alison Pollack)

Cribraria minutissima Schwein. (Figure 8-25)

Cribraria minutissima is the same size as *C. confusa* (page 154, Figure 8-24) but is darker in color and the lower half of the peridial net is represented by an urn-shaped calyculus. Like *C. confusa*, *C. minutissima* is most likely to be encountered on bark from living trees placed in moist chamber cultures.

Figure 8-25. *Cribraria minutissima* (©Yuri Novozholov)

Cribraria rufa (Roth) Rostaf.

(Figure 8-26)

This species can be recognized by the combination of the bright orange color of the fruiting bodies, a peridial net without distinct nodes, and the black stalk. *Cribraria rufa* occurs on decaying wood, almost exclusively wood from conifers. The species is widespread but not particularly common.

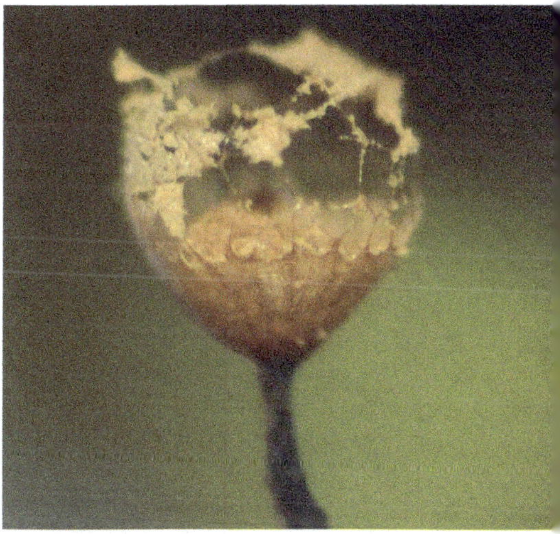

Figure 8-26. *Cribraria rufa* (©Alain Michaud)

Diachea subsessilis Peck

(Figure 8-27)

Diachea subsessilis shares some features (the white stalk and iridescent peridium) in common with *D. leucopodia* (page 93), but the sporotheca is globose and not cylindrical or elliptical as is the case for the latter species. Moreover, the spores of *D. subsessilis* are reticulate as opposed to warted in *D. leucopodia*. *Diachea subsessilis* typically occurs on leaf litter and other types of plant debris but sometimes fruits on living plants.

Figure 8-27. *Diachea subsessilis* (©Alison Pollack)

Figure 8-28. *Diderma fallax* (©Alain Michaud)

Figure 8-29. *Diderma hemisphaericum* (©Jorge Almeida)

Figure 8-30. *Diderma montanum* (©Alex Michaud)

Diderma fallax (Rostaf.) E. Sheld.

(Figure 8-28)

In earlier treatments of the myxomycetes, this species is listed as *Diderma lyalli* (Massee) Macbr. The fruiting body has a subglobose sporotheca, a prominent columella, and large (14 to 17 µm) coarsely reticulate spores. This is a species of alpine habitats, where it occurs on plant debris and low-growing living plants at the edge of melting snowbanks. On the basis of recent molecular data, *Diderma fallax* was placed in the genus *Polyschismium* as *P. fallax* (Rostaf.) A. Ronikier, J.M. García-Martín, A. Kuhnt, J.C. Zamora, M. de Haan, Janik & Lado.

Diderma hemisphaericum (Bull.) Hornem.

(Figure 8-29)

The fruiting bodies of *Diderma hemisphaericum* resemble miniature mushrooms. The discoid sporotheca is white, smooth, and flat on top with no evidence of the lime crystals that are present on the peridium of the morphologically similar *Didymium clavus* (page 158, Figure 8-34). *Diderma hemispaericum* occurs on leaf litter and various types of plant debris.

Diderma montanum (Meyl.) Meyl.

(Figure 8-30)

The fruiting body is a short-stalked sporangium with a globose, slightly depressed sporotheca. The peridium is smooth or slightly roughened and pinkish gray to nearly white. The single most distinctive feature is the prominent white or reddish brown club-shaped columella. *Diderma montanum* occurs on leaf litter and other types of plant debris. There are some reports of this species from alpine habitats, but the majority of records are from lowland habitats.

Diderma radiatum (L.) Morgan

(Figure 8-31)

The globose sporangia of *Diderma radiatum* occur on decaying wood. When dehiscence takes place the outer layer of the peridium splits apart into lobes that expand outward like the petals on a flower. This feature makes *D. radiatum* one of the more distinctive myxomycetes. This species is not particularly common and is restricted largely to temperate regions of the world

Figure 8-31. *Diderma radiatum* (©Henri Koskinen)

Diderma spumarioides (Fr. & Palmquist) Fr.

(Figure 8-32)

The hypothallus is not apparent or relatively inconspicuous in most myxomycetes. However, in *Diderma spumarioides*, it is the most obvious feature of the fruiting bodies, which are deeply embedded in a thick white hypothallus. Fruiting bodies of this species typically occur on leaf litter and various types of plant debris, but it is not unusual to find them on low-growing living plants. Some fruitings can be rather extensive.

Figure 8-32. *Diderma spumarioides* (©Alex Michaud)

Didymium anellus Morgan

(Figure 8-33)

The fruiting bodies of *Didymium anellus* are sporangia or short plasmodiocarps, with gray lime and circumscissile dehiscence. Fruiting bodies are ring-shaped and depressed in the center. This species occurs on leaf litter in nature or on bark from living trees placed in moist chamber cultures.

Figure 8-33. *Didymium anellus* (©Juan Francisco Moreno)

Figure 8-34. *Didymium clavus* (©Alain Michaud)

Figure 8-35. *Didymium difforme* (©Juan Francisco Moreno)

Figure 8-36. *Didymium dubium* (©Alain Michaud)

Didymium clavus (Alb. & Schwein.) Rabenh.

(Figure 8-34)

The fruiting bodies of *Didymium clavus* consist of a discoid sporotheca on a relatively short stalk. This species is morphologically similar to *Diderma hemisphericum* (page 156, Figure 8-29), but the surface of the peridium in the two species is different (crystalline lime in the former and a compact layer of lime granules in the latter). Moreover, the sporotheca of *D. clavus* is clearly umbilicate. This species occurs on decaying wood or various types of plant debris. It appears to be relatively uncommon.

Didymium difforme (Pers.) Gray

(Figure 8-35)

Didymium difforme is one of the members of the genus *Didymium* in which the peridium is eggshell-like and thus superficially similar to the peridium characteristic of species of *Diderma*. Although the crystalline nature of lime making up the peridium of *D. difforme* can be discerned with a good hand lens, the easiest way to identify this species in the field is to disrupt the outer limy layer of the peridium to reveal the pale purple iridescent inner peridium. The fruiting bodies of *D. difforme* are pulvinate or sometimes plasmodiocarpous. This species occurs on all types of plant debris and is often common on dung.

Didymium dubium Rostaf.

(Figure 8-36)

The fruiting bodies *Didymium dubium* are somewhat effused plasmodiocarps that can reach a length of 15 mm and a width of 7 mm. The most distinctive features are the very large stellate lime crystals on the peridium and a branched, very elastic capillitium that forms an open net. *Didymium dubium* is found on all types of plant debris. This species is considered to be primarily alpine in distribution, but specimens also have been recorded from lowland habitats.

Didymium nigripes (Link) Fr.

(Figure 8-37)

Didymium nigripes is very similar in appearance to *D. iridis* (page 101) but has a dark brown columella and a dark brown to almost black stalk. In *Didymium iridis*, the columella is white and the stalk is yellowish brown. Like the latter species, *D. nigripes* commonly occurs on leaf litter and various types of plant debris.

Figure 8-37. *Didymium nigripes* (©Alain Michaud)

Didymium squamulosum (Alb. & Schwein.) Fr.

(Figure 8-38)

Like other members of the genus of *Didymium*, the fruiting bodies of *D. squamulosum* have a peridium covered with stellate white lime crystals. It can be distinguished from other common species of *Didymium* by the relatively short and stout white stalk and the often "crinkly" appearance of the peridium. The usual substrates for *D. squamulosum* are leaf litter and other types of plant debris. This species is common on aerial litter in tropical forests.

Figure 8-38. *Didymium squamulosum* (©Paul Bos)

Didymium vaccinum (Durieu & Mont.) Buchet

(Figure 8-39)

Didymium vaccinum, like *D. difforme* (page 158, Figure 8-35), has a peridium that causes it to resemble a species of *Diderma*. The fruiting bodies are distinctive, especially after dehiscence. They are sessile or have a short stalk, and the sporotheca has the shape of a hemisphere. After the peridium falls away at dehiscence, it leaves behind a cup-like structure with a conspicuous columella arising from the middle. *Didymium* is common in deserts, where it fruits on old dead portions of cacti and other succulents.

Figure 8-39. *Didymium vaccinum* (©Yuri Novozholov)

Figure 8-40. *Elaeomyxa cerifera* (©Steve Young)

Elaeomyxa cerifera (G. Lister) Hagelst.

(Figure 8-40)

Elaeomyxa certifera resembles a member of the genus *Diachea* but differs in the presence of oils or waxes in the stalk, columella, or capillitium. This is usually very apparent at the base of the globose sporotheca and is a feature not shared with any other myxomycete. *Elaeomyxa cerifera* is found on moist decaying wood, usually along with bryophytes, in moist montane forests. It is not common.

Figure 8-41. *Hemitrichia clavata* (©Alain Michaud)

Hemitrichia clavata (Pers.) Rostaf.

(Figure 8-41)

Hemitrichia clavata is easily confused with *H. calyculata* (page 106, Figure 7-25) but has a much deeper, vase-like calyculus that gradually tapers downward to merge into the short, thick stalk. Moreover, the sporangia produced by *H. clavata* tend to be crowded closely together, while those of *H. calyculata* tend to be scattered. *Hemitrichia clavata* could be confused with *Trichia decipiens* (page 141) but has a very different capillitium.

Figure 8-42. *Lepidoderma chailletii* (©Alain Michaud)

Lepidoderma chailletii Rostaf.

(Figure 8-42)

In members of the genus *Lepidoderma*, the lime present on the peridium occurs in the form of thin plates. These can be scattered, with gaps between them, or closely aggregated as a distinct layer. The latter is the case for *Lepidoderma chailletii*. The fruiting body of this species is a sessile subglobose or pulvinate sporangium, 0.5 to 1.0 mm in diameter, and occurring in closely aggregated groups. This species occurs on low-growing living plants and plant debris at the edges of melting snowbanks. Based on recent molecular data, *Lepidoderma chailletii* was placed in the genus *Polyschismium* as *P. chailletii* (Rostaf.) A. Ronikier, J.M. García-Martín, A. Kuhnt, J.C. Zamora, M. de Haan, Janik & Lado.

Lepidoderma trevelyanii (Grev.) Poulain & Mar. Mey.

(Figure 8-43)

In earlier treatments of the myxomycetes, this species is listed as *Diderma trevelyanii* (Grev.) Fr. The fruiting body is a stalked sporangium with a pale reddish brown peridium marked with pale lines of dehiscence. At first the sporotheca is globose, but with dehiscence the peridium splits into stellate lobes. *Lepidoderma trevelyanii* occurs in alpine areas, where the fruiting bodies occur on plant debris and low-growing living plants at the edge of melting snowbanks. On the basis of recent molecular data, *Lepidoderma trevelyanii* was placed in the genus *Polyschismium* as *P. trevelyanii* (Rostaf.) A. Ronikier, J.M. García-Martín, A. Kuhnt, J.C. Zamora, M. de Haan, Janik & Lado.

Figure 8-43. *Lepidoderma trevelyanii* (©Yuri Novozhilov)

Licea succulenticola Mosquera, Lado, Estrada & Beltrán-Tej.

(Figure 8-44)

Some species of myxomycetes are possible to identify even when they possess few distinctive morphological features. This is true for *Licea succulenticola*, in which the subglobose fruiting bodies are small and easily overlooked in the field. However, this species usually can be identified on the basis of substrate, since it is restricted largely or completely to one specific type of substrate—the dead portions of cacti and other succulent plants in deserts.

Figure 8-44. *Licea succulenticola* (©Alison Pollack)

Lycogala conicum Pers.

(Figure 8-45)

Members of the genus *Lycogala* generally produce fruiting bodies (aethalia) that are more-or-less globose. The exception is *Lycogala conicum*, in which the aethalium is clearly conical. This species occurs on decaying wood and is not especially common.

Figure 8-45. *Lycogala conicum* (©Juan Francisco Moreno)

Figure 8-46. *Lycogala flavofuscum* (©Emily Johnson)

Lycogala flavofuscum (Ehrenb.) Rostaf.

(Figure 8-46)

This is the largest species in the genus *Lycogala*, with fruiting bodies (aethalia) capable of reaching 2 to 4 cm in total extent. Because of its large size and relatively thick cortex, *Lycogala flavofuscum* could be mistaken for a small puffball! This species occurs on decaying wood and sometimes on dead standing trees some distance above the ground.

Figure 8-47. *Macbrideola cornea* (©Alain Michaud)

Macbrideola cornea (G. Lister & Cran) Alexop.

(Figure 8-47)

Bark from living trees placed in moist chamber cultures often will yield very small myxomycetes that have stalked fruiting bodies with a globose sporotheca. At one time, those examples with dark spores were assigned to the genus *Comatricha*, but more modern taxonomic treatments have placed them in a number of other genera, including *Macbrideola* and *Paradiacheopsis*. *Macbrideola cornea* is one member of this assemblage of myxomycetes. The total height of the fruiting body is no more than about 2 mm. The most distinctive feature (shared with other members of the genus *Macbrideola*) is a stalk that is hollow. Because of the minute size of the fruiting body, this and other features have to be studied with the aid of a microscope to identify a particular specimen. *Macbrideola cornea* occurs on bark from living trees placed in moist chamber cultures.

Figure 8-48. *Meriderma carestiae* (©Alain Michaud)

Meriderma carestiae (Ces. & De Not.) Mar. Mey. & Poulain

(Figure 8-48)

In earlier treatments of the myxomycetes, this species is listed as *Lamproderma carestiae* (Ces & De Not.) Meyl., but it is now placed in the genus *Meriderma* because of the presence of a capillitium with funnel-shaped ends (page 65, Figure 6-32). The distinguishing features of the fruiting bodies of *M. carestiae* are their relatively large size (up to 2.5 mm tall), the subglobose to broadly ovate sporotheca, the short stalk, and the violet-blue or bonze iridescent peridium. This is one of the most common species in alpine habitats where it fruits on ground litter and low-growing plants at the edge of the melting snowbanks.

Meriderma cribrarioides (Fr.) Mar. Mey. & Poulain

(Figure 8-49)

Meriderma cribrarioides, which is listed in earlier treatment of the myxomycetes as *Lamproderma cribrarioides* (Fr.) R.E. Fr., is very similar in appearance to *M. carestiae* and also occurs on plant debris and living plants in alpine snowbank habitats. The fruiting bodies of the former are slightly smaller (no more than about 1.5 mm tall versus 2 mm in *M. carestiae*), but the main distinguishing feature of *M. cribrarioides* is the ornamentation on the spores, which is strongly reticulate.

Figure 8-49. *Meriderma cribrarioides* (©Alain Michaud)

Metatrichia floriformis (Schwein.) Nann.-Bremek.

(Figure 8-50)

In most earlier treatments of the myxomycetes, this species is listed as *Trichia floriformis* (Schwein.) G. Lister. It is easy to recognize from the clustered dark red to reddish black, globose sporangium-like units on a common stalk. The peridium dehisces into a series of small lobes that fold back to reveal the orange spore mass. *Metatrichia floriformis* occurs on decaying wood, often in extensive fruitings.

Figure 8-50. *Metatrichia floriformis* (©Marius Muessler)

Oligonema flavidum (Peck) Peck

(Figure 8-51)

This species could be confused with a members of the genus *Trichia* since the fruiting bodies are sporangia and the spores are yellow. For example, it is generally similar to *Trichia favoginea* (page 142), which shares the characteristic of having fruiting bodies that occur in dense clusters. However, the elaters that make up the capillitium of *Oligonema flavidum* are different from those of any species of *Trichia* since there are no conspicuous spiral bands. This species is associated with twigs and other woody debris but is not particularly common.

Figure 8-51. *Oligonema flavidum* (©Alain Michaud)

Figure 8-52. *Paradiacheopsis solitaria* (©Alain Michaud)

Figure 8-53. *Perichaena chrysosperma* (©Dmytro Leontyev)

Figure 8-54. *Perichaena corticalis* (©Kim Fleming)

Paradiacheopsis solitaria (Nann.-Bremek.) Nann.-Bremek.

(Figure 8-52)

The fruiting body of *Paradiacheopsis solitaria* is usually no more than about 0.5 mm tall and consists of a globose sporotheca on a slender black stalk that has a fibrous, netted base. The spores are dark brown and relatively large (14 to 16 μm), and the branches of the capillitium radiate outward from near the apex of the stalk. Like other minute myxomycetes, a microscope is required for an accurate identification. *Paradiacheopsis solitaria* appears on bark from living trees placed in moist chamber cultures.

Perichaena chrysosperma (Curr.) Lister

(Figure 8-53)

The fruiting bodies of *Perichaena chrysoperma* are typically short plasmodiocarps, and sometimes these are ring-shaped (looking like a miniature donut). When the latter is the case, little else is required to identify this species. However, when these are not present, another distinctive feature is the presence of a capillitium with prominent spines. This species occurs on decaying wood and bark in nature but is more likely to be encountered on bark from living trees placed in moist chamber cultures. It was recently determined that the original name proposed for this species and thus the correct one to use is *Ophiotheca chrysosperma* Curr.

Perichaena corticalis (Batsch) Rostaf.

(Figure 8-54)

The fruiting bodies of *Perichaena corticalis* tend to occur in tightly packed clusters on the substrate upon which they are found. The distinguishing features of the individual fruiting bodies are their hemispherical to slightly flattened shape, the bright reddish brown color of the peridium, and the distinct equatorial line of dehiscence that divides the peridium into upper and lower portions. *Perichaena corticalis* typically occurs on decaying wood and dead bark.

Perichaena dictyonema Rammeloo

(Figure 8-55)

The fruiting bodies of *Perichaena dictyonema* are pillow-shaped or dome-shaped and thus somewhat resemble those of *P. depressa* (page 123). However, the fruiting bodies of *P. dictyonema* are scattered and not clustered. Moreover, the membranous inner peridium is prominently exposed as a wide yellow band around the middle of the fruiting body, dividing the latter into upper and lower portions. This species is most likely to be encountered on twigs in moist chamber cultures.

Figure 8-55. *Perichaena dictyonema* (©Alison Pollack)

Physarum bitectum G. Lister

(Figure 8-56)

Physarum bitectum can appear to be very similar to *P. bivalve* (page 127) when the latter forms short plasmodiocarps. Both species typically occur on leaf litter, but the fruiting bodies of *P. bivalve* tend to be more laterally compressed while those of *P. bitectum* are somewhat rounded in cross section. Moreover, the spores of *P. bitectum* are ornamented with coarse spines in contrast to the finely warted spores of *P. bivalve*.

Figure 8-56. *Physarum bitectum* (©Alain Michaud)

Physarum compressum Alb. & Schwein.

(Figure 8-57)

This is one of the more easily recognized members of the genus *Physarum*. The sporotheca is characteristically fan-shaped or kidney-shaped with a white peridium. Since *Physarum compressum* is rather variole, other shapes are possible. However, a least some typical examples usually occur in any fruiting. The usual substrate is leaf litter and other types of plant debris, but the species also occurs on dung. In tropical forests, *P. compressum* often forms extensive fruitings.

Figure 8-57. *Physarum compressum* (©Alison Pollack)

Figure 8-58. *Physarum crateriforme* (©Juan Francisco Moreno)

Figure 8-59. *Physarum decipiens* (©Juan Francisco Moreno)

Figure 8-60. *Physarum didermoides* (©Laurie Leonard)

Physarum crateriforme Petch

(Figure 8-58)

The fruiting bodies of *Physarum crateriforme* can be tentatively identified by the club-shaped or goblet-shaped sporotheca with a grayish white to very pale brown peridium and the black stalk. However, the single most distinguishing feature is the presence of a prominent calcareous columella that extends almost to the apex of the sporotheca. Fruiting bodies are usually scattered across the substrate upon which they occur. This species occurs on the bark of living trees, including bark placed in moist chamber cultures.

Physarum decipiens M.A. Curtis

(Figure 8-59)

There are a number of species of *Physarum* which produce sessile fruiting bodies that are either sporangia or short to branched plasmodiocarps. *Physarum decipiens* is one of the more common species occurring on bark from living trees placed in moist chamber cultures. The distinguishing features are the yellow "bumpy-looking" (due to the presence of prominent lime scales) peridium and the large lime nodes of the capillitium.

Physarum didermoides (Pers.) Rostaf.

(Figure 8-60)

The fruiting bodies of *Physarum didermoides* occur in dense clusters on a prominent white limy hypothallus. The distinctive features of the fruiting bodies are the short weak stalk and a cylindrical or ovoid sporotheca with a peridium in which the outermost of two layers quickly falls away. It is possible that this species could be confused with *Diderma spumarioides* (page 157, Figure 8-32), but in the latter species the individual fruiting bodies have a smooth peridium and are immersed in the hypothallus. *Physarum didermoides* occurs on the dead bark of logs and on various types of plant debris. It was recently determined that the original name proposed for this species and thus the correct one to use is *Claustria didermoides* (Pers.) Fr.

Physarum echinosporum Lister

(Figure 8-61)

The fruiting bodies of this species are elongated laterally compressed plasmodiocarps with a smooth white peridium. The general aspect of *Physarum echinosporum* is similar to that of *P. bivalve* (page 127), but the peridium of the latter species is never as smooth or as pure white. However, the single most important difference is the very spiny spore of *P. echinosporum*. The spore of *Physarum bivalve* is finely warted. *Physarum echinosporum* has a tropical distribution but occasionally appears on bark from living trees placed in moist chamber cultures.

Figure 8-61. *Physarum echinosporum* (©Alain Michaud)

Physarum melleum (Berk. & Broome) Massee

(Figure 8-62)

This species can be recognized from the combination of the relatively stout white stalk, the yellow to dull orange peridium, and the large white lime nodes of the capillitium. The color of the peridium can fade in older specimens, but the overall morphology of the fruiting body is distinctive enough to suggest this species. *Physarum melleum* usually fruits on leaf litter and other types of plant debris. On the basis of recent molecular data, *Physarum melleum* was placed in the new genus *Nannengaella* as *N. mellea* (Berk. & Broome) J.M. García-Martín, J.C. Zamora & Lado.

Figure 8-62. *Physarum melleum* (©Joe Takano)

Physarum pusillum (Berk. & M.A. Curtis) G. Lister

(Figure 8-63)

Figure 8-63. *Physarum pusillum* (©Yuri Novozhilov)

The fruiting body *Physarum pusillum* consists of a relatively long stalk (more than half of the total height) and a white and erect globose sporotheca with a reddish brown base. The latter is less fragile than the upper portion of the peridium and persists after dehiscence. As such, an older specimen of *P. pusillum* might suggest that it is a species of *Craterium*, but no member of the latter genus has as long a stalk. *Physarum pusillum* is usually associated with leaf litter but can occur on the bark from living trees placed in moist chamber cultures.

Figure 8-64. *Physarum roseum* (©Alison Pollack)

Figure 8-65. *Physarum viride* (©Henri Koskinen)

Figure 8-66. *Reticularia lycoperdon* (©Henri Koskinen)

Physarum roseum Berk. & Broome

(Figure 8-64)

There are several species of *Physarum* that produce stalked fruiting bodies with a globose to ovoid sporotheca. *Physarum roseum* is distinctive because of the bright red color of the peridium. Interestingly, the plasmodium is also red, which makes this species one of the few myxomycetes that can be reliably identified at this stage of the life cycle. *Physarum roseum* occurs on dead bark and leaf litter. Based on recent molecular data, *P. roseum* was placed in the genus *Craterium* as *C. roseum* (Berk. & Broome) J.M. García-Martín & Lado.

Physarum viride (Bull.) Pers.

(Figure 8-65)

Physarum viride is one of the more common species of *Physarum*. The fruiting body is a stalked sporangium with a lens-shaped, typically nodding sporotheca. This species is very similar in appearance to *P. album* (page 126), but the latter has a white peridium while that of *P. viride* is bright yellow or sometimes orange. *Physarum viride* occurs on decaying wood and dead bark, although it is also not uncommon on bark from living trees placed in moist chamber cultures.

Reticularia lycoperdon Bull.

(Figure 8-66)

Reticularia lycoperdon is one of the largest myxomycetes, producing fruiting bodies (aethalia) that can reach 10 cm in total extent. At first these have a silvery white cortex, but when this is disturbed, the cocoa-brown spores are revealed. This species typically occurs on the trunks of standing dead trees, sometimes several feet above the ground. The fruiting bodies are often invaded by slime mold beetles, whose presence is indicated by their entry holes in the cortex.

Stemonaria longa (Peck) Nann.-Bremek., R. Sharma, & Y. Yamam.

(Figure 8-67)

This species is listed as *Comatricha longa* Peck in most earlier treatments of the myxomycetes. The fruiting bodies have the general appearance of those found in a member of the genus *Stemonitis*, but they are drooping and not erect. The capillitium is distinctive, consisting of straight or forked branches that extend outward at a 90-degree angle from the columella. This is a species found on decaying wood in tropical forests.

Figure 8-67. *Stemonaria longa* (©Laurie Leonald)

Stemonitis splendens Rostaf.

(Figure 8-68)

This species is largest member of the genus *Stemonitis*, producing tuffs of fruiting bodies that can be as much as 15 mm tall. The reddish brown color of the spores does not distinguish this species from several other members of the genus, but none of the others have the very large meshes in the surface net. *Stemonitis splendens* occurs on decaying wood, often as extensive fruitings.

Figure 8-68. *Stemonitis splendens* (©Alain Michaud)

Trichia botrytis (J.F. Gmel.) Pers.

(Figure 8-69)

Trichia botrytis is one of several stalked members of the genus *Trichia* that produce fruiting bodies with a globose to somewhat pear-shaped sporotheca. The distinguishing features are the elaters with long slender tips, the reddish brown to black color of the peridium, and the presence of yellow lines of dehiscence on the apex of the sporotheca. The fruiting bodies are relatively large (2 to 4 mm total height) and occur on decaying wood.

Figure 8-69. *Trichia botrytis* (©Alain Michaud)

Figure 8-70. *Trichia contorta* (©Alain Michaud)

Figure 8-71. *Trichia scabra* (©Alain Michaud)

Figure 8-72. *Trichia subfusca* (©Alain Michaud)

Trichia contorta (Ditmar) Rostaf.

(Figure 8-70)

Trichia contorta is very variable (several different varieties have been recognized) and might require other possible species of *Trichia* to be eliminated to reach a positive identification. The fruiting bodies are sporangia or short plasmodiocarps with a cartilaginous dull yellow-brown to dark reddish brown peridium, bright ochraceous warted spores, and elaters in which the spirals are often decorated with short spines. This species usually occurs on dead bark in nature and sometimes appears on bark from living trees placed in moist chamber cultures.

Trichia scabra Rostaf.

(Figure 8-71)

The densely clustered fruiting bodies of *Trichia scabra* resemble those of *T. favoginea* (page 142) but those of the latter have a yellow peridium. In *T. scabra*, the peridium is dull orange or golden brown. Moreover, the latter species has a well-developed dark hypothallus that is lacking in *T. favoginea*. *Trichia scabra* occurs on decaying wood or bark and is less common than *T. favoginea*.

Trichia subfusca Rex

(Figure 8-72)

Trichia subfusca is similar in appearance to *T. botrytis* (page 169, Figure 8-69) but has a darker peridium (dark brown to almost black), is smaller (total height 0.2 mm as compared to 5.0 mm for the latter species), and has larger spores (11 to 15 µm versus 9 to 10 µm). *Trichia subfusca* often occurs as solitary fruiting bodies on decaying wood or sometimes on mosses.

Trichia verrucosa Berk.

(Figure 8-73)

The distinguishing features of *Trichia verrucosa* are the clusters of sporangium-like units on united stalks, the bright yellow to ochraceous peridium, and spores that are coarsely reticulate (similar to those of *Trichia favoginea* (page 142). *Trichia verrucosa* occurs on decaying wood, often in extensive fruitings. Interestingly, this species is common in the Southern Hemisphere but rare in the Northern Hemisphere. Based on recent molecular data, *Trichia verrucosa* was placed in the genus *Oligonema* as *O. verrucosum* (Berk.) García-Cunch., J.C. Zamora & Lado.

Figure 8-73. *Trichia verrucosa* (©Steve Young)

Willkommlangea reticulata (Alb. & Schwein.) Kuntz.

(Figure 8-74)

In earlier treatments of the myxomycetes this species is listed as *Cienkowskia reticulata* (Alb. & Schwein.) Rostaf. *Willkommlangea reticulata* is one of the easiest myxomycetes to identify. The worm-like plasmodiocarps have an orange peridium on which there are conspicuous red dots, a combination of features that is unique to *W. reticulata*. This species occurs on decaying wood, especially on twigs and small branches. It also sometimes appears on the bark from living trees placed in moist chamber cultures.

Figure 8-74. *Willkommlangea reticulata* (©Yura Novozhilov)

Glossary

A

aethalium (plural: aethalia): a relatively large, sessile, round or mound-shaped fruiting body formed from all or a major portion of a plasmodium

amoeboflagellate cell: a general term used to refer to either of the uninucleate haploid trophic stages (i.e., myxamoeba or swarm cell) in the myxomycete life cycle

anastomosed: interconnected or joined repeatedly to form a network

angular: having an angle or angles; not rounded in outline

aphanoplasmodium (plural: aphanoplasmodia): a type of plasmodium that is flat, transparent, and difficult to observe in nature; it generally is evident only when it heaps up in preparation for the formation of fruiting bodies; characteristic of members of the traditional order Stemonitales

apomictic: a type of life cycle in which meiosis and the fusion of gametes do not occur; as a result, all stages in the life cycle are diploid

areolate: having a surface divided into smaller areas by cracks or crevices

asperulate: having a surface that appears rough because of the presence of small warts or spines

aerial litter microhabitat: dead but still attached plant parts above the ground

attenuate: gradually narrowed; becoming smaller and thinner

B

badhamioid: a type of capillitium consisting entirely of limy elements

binary fission: a type of asexual reproduction in which a cell divides into two equal, or nearly equal, daughter cells

binomial: a two-word term that represents the name of a particular species

bryophilous: living on, or at least associated with, bryophytes (mosses and liverworts)

C

calyculus (plural: calyculi): a persistent cuplike structure formed from the basal portion of the peridium and surrounding the base of the spore mass

capillitium (plural: capillitia): a system of sterile elements found within the spore mass of many myxomycetes

cartilaginous: consisting of a stout, uniformly thickened and usually rather stiff layer

cellular slime molds: a group of organisms with a uninucleate, amoeboid trophic stage similar to the one found in the myxomycetes but not forming plasmodia or the types of fruiting bodies characteristic of the latter group; also called dictyostelids

cinereous: bluish gray; the color of ashes

circumscissile: dehiscing along a line around the middle

clade: a group of organisms that share a common ancestor

clavate: club-shaped

clustered: occurring close together in the same small area of the substrate

columella (plural: columellae): a sterile structure that extends into the spore mass from below; in stipitate forms it appears to represent an extension of the stalk that continues upward

compressed: pressed together; flattened

conical: more or less cone-shaped

convoluted: coiled or twisted

coprophilous: dung-inhabiting; occurring on dung

cortex: the thick covering over the spore mass of an aethalium

corticolous: living on bark; associated with the bark surface of living trees

cylindrical: having the general form of a cylinder; a term applied to a structure with essentially the same diameter throughout its length

D

dehiscence: the splitting of the peridium (or cortex) in mature fruiting bodies to expose the spore mass

dichotomous: branching or forking into two more or less equal parts

dictydine granules: microscopic, usually dark-colored structures found in the fruiting bodies of some members of the traditional order Liceales; also called plasmodic granules

diploid: the condition in which a nucleus has two sets of chromosomes (2n) as opposed to only a single set (n); it is established in the zygote following karyogamy

discoid: disk-shaped

dictyostelids: *see* cellular slime molds

E

echinate: spiny; having a surface covered with sharp-pointed spines

ecology: the study of the interrelations of living things and their environment

effused: flattened or spread out over the substrate

elater: a single, free, simple or branched capillitial strand; characteristic of some members of the traditional order Trichiales

epiphytic: growing or occurring upon plants

F

fasciculate: occurring close together in small groups or bundles

ferruginous: rust-colored; having the color of iron rust

flagellate: possessing one or more flagella

flagellum (plural: flagella): a relatively long, whiplike structure used in locomotion; characteristic of swarm cells

flexuous: having bends or curves; crooked

fruiting body: a general term for the spore-producing structure produced during the reproductive stage in the myxomycete life cycle; the structure within which (or in one case, on the surface of which) the spores are produced; also referred to as a fructification or sporophore

fungivorous: fungus-feeding; feeding upon the fruiting bodies or vegetative hyphae of fungi

fuscous: brownish gray; smoky drab or dusky in color

fusiform: spindle-shaped; tapering at both ends

G

gamete: a haploid reproductive cell capable of fusing with another similar cell to produce a diploid zygote

globose: spherical; having the general form or shape of a ball

gregarious: occurring relatively close together in the same general area of the substrate but not clustered

H

Hantsch's fluid: a mounting medium used in the microscopic examination of fruiting bodies

haploid: the condition in which a nucleus has only one set of chromosomes; it exists in cells that can potentially function as gametes

hyaline: clear or transparent

hygroscopic: absorbing or releasing moisture in response to changes in atmospheric humidity and consequently changing shape

hypha (plural: hyphae): one of the individual strands or threadlike elements that make up the vegetative body of a fungus

hypothallus (plural: hypothalli): a thin layer deposited by a plasmodium at the time of fruiting; located on the substrate beneath one or more fruiting bodies

I

iridescent: displaying rainbow-like colors

K

karyogamy: the fusion of two nuclei following plasmogamy, which results in a diploid zygote

L

lactophenol: a mounting medium used in the microscopic examination of fruiting bodies

lenticular: shaped like a double convex lens

life cycle: the series of events or stages involved in the perpetuation of an organism

lignicolous: living on or within wood

lime node: an enlarged, lime-containing junction in the capillitium; characteristic of some members of the traditional order Physarales

limy: containing lime (calcium carbonate)

M

macrocyst: a cell-like unit making up a sclerotium

meiosis: reduction division; a type of nuclear division in which a single diploid cell forms four haploid nuclei

membranous: consisting of a thin, pliable layer

microcyst: a resistant structure formed when a myxamoeba ceases its activity and becomes dormant

mitosis: nuclear division characterized by the formation of two identical daughter nuclei

morphology: the study of the form or structure of an organism

morphospecies: a concept in which a species is recognized solely on the basis of morphological features

myxamoeba (plural: myxamoebae): the microscopic, uninucleate amoeboid trophic stage in the myxomycete life cycle that also may function as a gamete

myxomyceticolous: living on or within the fruiting bodies of myxomycetes

N

node: an expanded junction in the peridial net of the fruiting body in some members of the traditional order Liceales

O

obovate: ovate with the narrow end at the base

obpyriform: pear-shaped with the broad end uppermost

ochraceous: pale yellow to reddish yellow

olivaceous: dull yellowish green; the color of an olive

operculum: a lid or cover

ovate: egg-shaped; having the general form or shape of an egg

ovoid: egg-shaped; usually pertaining to solid bodies

P

palisade: a condition in which more or less elongated fruiting bodies or other structures are closely packed together in a perpendicular arrangement

pallid: having a pale or white appearance

peridial net: a reticulate structure formed when certain portions of the peridium persist in a mature fruiting body; characteristic of some members of the traditional order Liceales

peridium (plural: peridia): the covering over the spore mass of a fruiting body

persistent: retaining the original shape or structure; not breaking up or disappearing

petaloid: having a pattern of dehiscence in which the peridium splits into lobes arranged like the petals of a flower; also called floriform

phaneroplasmodia (plural: phaneroplasmodium): a type of plasmodium that may be relatively large and conspicuous enough to be observed in nature; it is characterized by a fan-shaped advancing margin

physaroid: a type of capillitium consisting of slender non-calcareous threads connecting larger calcareous nodes

plasmodic granules: *see* dictydine granules

plasmodiocarp: a sessile, branched, ring-shaped, or netted type of fruiting body formed when a plasmodium becomes concentrated in its main veins (without breaking up into smaller units) during fruiting

plasmodiocarpous: producing (or at least suggestive of) the plasmodiocarp type of fruiting body

plasmodium (plural: plasmodia): the acellular, multinucleate mass of protoplasm representing the main trophic stage in the myxomycete life cycle

plasmogamy: the fusion of the protoplasts of two cells; usually followed by karyogamy

poroid: having more or less evident pores

prolate: elongated toward the poles

prostrate: lying flat on the substrate

protoplasmodium (plural: protoplasmodia): a type of plasmodium that remains microscopic, forms no veins, and gives rise to only a single rather small fruiting body

pseudoaethalium (plural: pseudoaethalia): a type of fruiting body that consists of a mass of sporangia tightly packed together to resemble an aethalium

pseudocapillitium (plural: pseudocapillitia): a system of irregular plates, tubes, or threadlike elements occurring within the spore mass of an aethalium and suggestive of a true capillitium but not formed in the same manner; characteristic of some members of the traditional order Liceales

pseudocolumella (plural: pseudocolumellae): a spherical to irregular lime mass more or less centrally located within the spore mass and consisting of a single discrete body or a loose aggregation of smaller units

pseudoplasmodium (plural: pseudoplasmodia): a multicellular mass produced during the life cycle of a cellular slime mold

pulverulent: having a surface that appears powdery

pulvinate: cushion-shaped; having the general shape of a small cushion

punctate: having a surface marked with very small warts or tiny depressions

pyriform: pear-shaped

R

recurved: curving downward or backward

reticulate: netted; having a surface covered with a network of spines, warts, or ridge

reticulum: a network of fine lines

rugose: wrinkled; having a surface marked by coarse wrinkles

S

saprophagous: feeding upon nonliving organic matter

sclerotium (plural: sclerotia): a dormant, hardened structure formed from the plasmodium under unfavorable environmental conditions

sessile: lacking a stalk; attached directly to the substrate

sinuous: wavy; having curves, bends, or turns

solitary: occurring alone on the substrate

sorocarp: the fruiting body produced by a cellular slime mold

sorophore: *see* stalk

spinulose: having a surface marked with small spines

spiny: *see* echinate

sporangiate: producing or suggestive of the sporangium type of fruiting body

sporangium (plural: sporangia): a type of fruiting body formed when a plasmodium breaks up into a number of small portions, each of which develops into a single stalked or sessile unit; a small plasmodium may form only one sporangium

spore ball: a group of spores joined closely together; characteristic of some members of the genus *Badhamia* and a few other myxomycetes

spore mass: all the spores within a single fruiting body when considered collectively

spore: a resistant microscopic reproductive structure produced in the fruiting body of a myxomycete

sporocarp: *see* fruiting body

sporotheca: the actual spore-containing portion of a sporangium

sporulate: to undergo sporulation

sporulation: the fruiting process in which a plasmodium is transformed into one or more fruiting bodies

stalk: a structure supporting the spore mass, which often is elevated some distance above the substrate; also called a stipe

stellate: star-shaped; having the general appearance of a star

stipe: *see* stalk

stipitate: stalked; having a definite stalk

striate: marked with fine parallel ridges

subcylindrical: somewhat cylindrical

subglobose: not quite globose

surface net: a well-defined, often highly branched and anastomosing peripheral network found just beneath the peridium, where it arises from the tips of the capillitial branches; characteristic of the genus *Stemonitis*

swarm cell: the microscopic, uninucleate flagellated trophic stage in the myxomycete life cycle that also may function as a gamete

T

terete: more or less round in cross section; cylindrical

trophic: pertaining to feeding or nutrition

truncate: square or broad at the end as if cut off

turbinate: top-shaped

U

umbilicate: having a depression or cavity at the base of the spore mass where the stalk is attached

V

venulose: having small veins

verrucose: warted; having a surface marked with warts

verruculose: minutely verrucose; very delicately warted

violaceous: bluish purple; the color of violets

W

warted: having a surface marked with small protuberances or raised areas

Z

zygote: a diploid cell formed as a result of the fusion of two gametes

References

Adamonyte, G. 2006. The genus *Echinostelium* (Myxomycetes) in *Lithuania. Acta Mycologica* 41: 169–176.

Adamonyte, G., S. L. Stephenson, A. Michaud, E.-H. Seraoui, M. Meyer, Y. K. Novozhilov, and T. Krivomaz. 2011. Myxomycete species diversity on the island of La Réunion (Indian Ocean). *Nova Hedwigia* 92: 523–549.

Aldrich, H. C., and J. W. Daniel, eds. 1982. *Cell Biology of Physarum* and *Didymium*, 2 vols. Academic Press, New York.

Alexopoulos, C. J. 1959. Myxomycetes from Greece. *Brittonia* 11: 25–40.

———. 1960. Gross morphology of the plasmodium and its possible significance in the relationships among the myxomycetes. *Mycologia* 52: 1–20.

———. 1963. The myxomycetes II. *Botanical Review* 29: 1–78.

———. 1964. The rapid sporulation of some myxomycetes in moist chamber culture. *Southwestern Naturalist* 9: 155–159.

———. 1965. First list of myxomycetes from Texas, with notes on some rarely reported species. *Southwestern Naturalist* 10: 221–226.

———. 1970. Rain forest myxomycetes. Pages F21–F23 in *A Tropical Rain Forest*, Odum, H. T., ed. Washington, DC: United States Atomic Energy Commission. Wiley & Sons, New York.

Alexopoulos, C. J., and M. R. Henney. 1971. Myxomycetes from Texas. II. Additions and corrections. *Southwestern Naturalist* 16: 143–150.

Alexopoulos, C. J., and J. A. Sáenz. 1975. The myxomycetes of Costa Rica. *Mycotaxon* 2: 223–271.

Angel, K., and D. T. Wicklow. 1975. Relationships between coprophilous fungi and fecal substrates in a Colorado grassland. *Mycologia* 67: 63–74.

Arambarri, A. M., and H. A. Spinedi. 1989. Antarctic Myxomycetes [Mixomicetes antarticos]. *Instituto Antartico Argentino*. Contribucion No. 365. Buenos Aires, Argentina.

Arenas-Taborda, A., M. C. García-Chaves, J. P. Niño-García, and C. Rojas. 2021. Myxomycete assemblage turnover across a moisture and elevation gradient in Costa Rica. *Slime Molds* V1A4: 1-14.

Barron, G. 1991. Protoplasm in motion. *Seasons* 31: 20–25.

Beltran, E., C. Lado, J. Barrera, and E. Gonzalez. 2004. Myxomycete diversity in the Laurel forest of Garajonay National Park (Canary Islands, Spain). *Systematics and Geography of Plants* 74: 159–173.

Binyamini, N. 1997. Myxomycetes from Israel IV. *Mycoscience* 38: 87–89.

Black, D. R., S. L. Stephenson, and C. A. Pearce. 2004. Myxomycetes associated with the aerial litter microhabitat in tropical forests of northern Queensland, Australia. *Systematics and Geography of Plants* 74: 129–132.

Blackwell, M. 1974. A study of the sporophore development in the myxomycete *Protophysarum phloiogenum*. *Archives of Microbiology* 99: 331–344.

———. 1984a. Distribution and sporulation phenology of myxomycetes in the Sonoran Desert of Arizona. *Microbial Ecology* 10: 369–377.

———. 1984b. Myxomycetes and their arthropod associates. Pages 67–90 in *Fungus-Insect Relationships: Perspectives in Ecology and Evolution*. Wheeler, Q. D., and M. Blackwell, eds. University Press, New York.

Blackwell, M., and R. L. Gilbertson. 1980. Sonoran Desert myxomycetes. *Mycotaxon* 11: 139–149.

———. 1984. Distribution and sporulation phenology of myxomycetes in the Sonoran Desert of Arizona. *Microbial Ecology* 10: 369–377.

Blackwell, M., T. Laman, and R. L. Gilbertson. 1982. Spore dispersal of *Fuligo septica* (Myxomycetes) by lathridiid beetles. *Mycotaxon* 14: 58–60.

Braun, K. L., and H. W. Keller. 1976. Myxomycetes of Mexico I. *Mycotaxon* 3: 297–317.

Brooks, T. E. 1967. *A Study of Corticolous Myxomycetes*. Ph.D. Dissertation, University of Kansas, Lawrence, Kansas.

Buxton, P. A. 1954. British Diptera associated with fungi. 2. Diptera bred from myxomycetes. *Proceedings of the Royal Entomological Society of London* (A) 29: 163–171.

———. 2009. British Diptera associated with fungi. 2.—Diptera bred from myxomycetes. *Physiological Entomology* 29: 163–171.

Buyck, B. 1988. The *Diderma spumarioides-globosum* complex (Myxomycetes). *Bulletin du Jardin Botanique Belgique* 58: 169–218.

Calaça, F. J. S., J. C. Araújа, G. Cacialli, N. C. Silva, C. Rojas, and S. Xavier-Santos. 2020. Fimicolous myxomycetes: overview of their global distribution and scientific production. *Biologia* 75: 2159–2174.

Clark, J., and O. R. Collins. 1976. Studies on the mating systems of eleven species of myxomycetes. *American Journal of Botany* 63: 783–789.

Clark. J., and E. F. Haskins. 2010. Reproductive systems in the myxomycetes: a review. *Mycosphere* 1: 337–353.

Coelho, I. L., and S. L. Stephenson. 2012. Myxomycetes associated with pipevine, a temperate liana. *Mycosphere* 3: 245–249.

Cohen, A. L. 1941. Nutrition of the myxomycetes. II. Relations between plasmodia, bacteria, and substrate in two-membered culture. *Botanical Gazette* 103: 205–224.

Collins, O. R. 1980. Apomictic-heterothallic conversion in a myxomycete, *Didymium iridis*. *Mycologia* 72: 1109–1116.

Cox, J. J. 1981. Notes on coprophilous myxomycetes from the western United States. *Mycologia* 73: 741–747.

Critchfield, R. L., and R. S. Demaree. 1991. Annotated checklist of California myxomycetes. *Madroño* 38: 45–56.

Crowder, W. 1926. Marvels of mycetozoa. *National Geographic* 49: 421–443.

Dagamac, N. H. A., M. Hoffman, Y. K. Novozhilov, and M. Schnittler. 2017. Myxomycetes from the highlands of Ethiopia. *Nova Hedwigia* 104: 111–127.

Dove, W. F., J. Dee, S. Hatano, F. B. Hugli, and K. E. Wohlfarth-Bottermann, eds. 1986. *The Molecular Biology of Physarum polycephalum*. Plenum Press, New York and London.

Eliasson, U. 1981. Patterns of occurrence of myxomycetes in a spruce forest in south Sweden. *Holarctic Ecology* 4: 20–31.

———. 1991. The myxomycete biota of the Hawaiian Islands. *Mycological Research* 95: 257–267.

———. 2013. Coprophilous myxomycetes: recent advances and future research directions. *Fungal Diversity* 59: 85–90.

Eliasson, U. H., H. W. Keller, and J. A. Hutchinson. 1988. Myxomycetes from Arkansas. *Mycotaxon* 32: 375–398.

Eliasson, U., and N. Lundqvist. 1979. Fimicolous myxomycetes. *Botaniska Notiser* 132: 551–568.

Ellis, M. B., and J. P. Ellis. 1988. *Microfungi on Miscellaneous Substrates: an Identification Handbook*. Timber Press, Portland, Oregon.

Emoto, Y. 1938. Untersuchungen uber die entwicklung der myxomyceten auf afulenden holzern. *Japanese Journal of Botany* 9: 253–257.

Emoto, Y. 1977. *The Myxomycetes of Japan*. Sangyo Tosho Publishing Company, Tokyo, Japan.

Estrada-Torres, A., D. Wrigley de Basanta, E. Conde, and C. Lado. 2009. Myxomycetes associated with dryland ecosystems of the Tehuacán-Cuicatlán Valley Biosphere Reserve, Mexico. *Fungal Diversity* 36: 17–56.

Evenson, A. E. 1961. A preliminary report of the myxomycetes of southern Arizona. *Mycologia* 53: 137–144.

Farr, M. L. 1958. Taxonomic studies in the Myxomycetes. I. The *Trichia favoginea* complex. *Mycologia* 50: 357–369.

———. 1968. An illustrated key to the myxomycetes of South America, with special reference to Brazil. *Rickia* 3: 43–88.

———. 1976. *Flora Neotropical Monograph No. 16. Myxomycetes.* The New York Botanical Garden, Bronx. New York.

———. 1981. *How to Know the True Slime Molds.* Wm. C. Brown Company, Dubuque, Iowa.

Farr, M. L., U. Eliasson, and K. P. Dupont. 1979. Myxomycetes from Ecuador. *Mycotaxon* 8: 127–134

Feest, A. 1987. The quantitative ecology of soil Mycetozoa. *Progress in Protistology* 2: 331–361.

Feest, A., and R. Campbell. 1986. The microbiology of soils under successive wheat crops in relation to take-all disease, *FEMS Microbiology Ecology* 38: 99-111.

Feest, A., and M. F. Madelin. 1985. A method for the enumeration of myxomycetes in soils and its application to a wide range of soils. *FEMS Microbiology Ecology* 31: 103–109.

———. 1988a. Seasonal population changes of myxomycetes and associated organisms in four woodland soils. *FEMS Microbiology Ecology* 53: 133–140.

———. 1988b. Seasonal population changes of myxomycetes and associated organisms in five non-woodland soils, and correlations between their numbers and soil characteristics. *FEMS Microbiology Ecology* 53: 141–152.

Fiore-Donno, A.-M., C. Berney, J. Pawlowski, and S. L. Baldauf. 2005. Higher-order phylogeny of plasmodial slime molds (Myxogastria) based on EF1A and SSU rRNA sequences. *Journal of Eukaryotic Microbiology* 52: 201–210.

Fiore-Donno, A.-M., F. Clissmann, M. Meyer, M. Schnittler, and T. Cavalier-Smith. 2013. Two-gene phylogeny of bright-spored Myxomycetes (slime moulds, superorder Lucisporidia). *PLoS ONE 8*, https://doi.org/10.1371/journal/pone.0062586

Fiore-Donno, A.-M., M. Meyer, S. L. Baldauf, and J. Pawlowski. 2008. Evolution of dark-spored Myxomycetes (slime-molds): molecules versus morphology. *Molecular Phylogenetics and Evolution* 46: 878–889.

Fiore-Donno, A.-M., S. I. Nikolaev, M. Nelson, J. Pawlowski, T. Cavalier-Smith, and S. L. Baldauf. 2010. Deep phylogeny and evolution of slime moulds (Mycetozoa). *Protist* 161: 55–70.

Fiore-Donno, A.-M., J. Weinert, T. Wubet, and M. Bonkowski. 2016. Metacommunity analysis of amoeboid protists in grassland soils. *Scientific Reports* 6: 19068 doi:10.1028/srep19068.

Frederick, L., I. L. Roth, and L. Pendergrass. 1986. Observations on the "peridial platelets" of *Clastoderma debaryanum. Mycologia* 78: 263–268.

Fukasawa, Y., K. Takahashi, T. Arikawa, T. Hattori, and N. Maekawa. 2015. Fungal wood decomposer activities influence community structures of myxomycetes and bryophytes on coarse woody debris. *Fungal Ecology* 14: 44–52.

Fullmer, E. H. 1921. The slime molds of Ohio. *Bulletin of the Ohio Biological Survey* 3: 1–72.

Gao, Y., X. Zhang, G. He, O. N. Shchepin, S. Yan, and S. Chen. 2019. Influence of forest type on dark-spored myxomycete community in subtropical forest soil, China. *Soil, Biology and Biochemistry* 139: doi.org/10.1016/j.soilbio.2019.107606.

García-Martín, J. M., J. C. Zamora, and C. Lado. 2023. Multigene phylogeny of the order Physarales (Myxomycetes, Amoebozoa): shedding light on the dark-spored clade. *Persoonia* 51:89-124.

Gilbert, F. A. 1928. Observation on the feeding habits of the swarm cells of myxomycetes. *American Journal of Botany* 15: 473–484.

Gilbert, H. C., and G. W. Martin. 1933. Myxomycetes found on the bark of living trees. *University of Iowa Studies in Natural History* 15: 3–8.

Goad, A. E., and S. L. Stephenson. 2013. Myxomycetes appearing in moist chamber cultures on four different types of leaves. *Mycosphere* 4: 707–712.

Gottsberger, G., and N. E. Nannenga-Bremelkamp. 1971. A new species of *Didymium* from Brazil. Proceedings of the Koninklijke Nederlandse Akademie van Wetenschappen, Series C, *Biological and Medial Sciences* 74: 264–268.

Gøtzsche, H. F. 1984. Contributions to the myxomycete flora of Iceland. *Acta Botanica Islandica* 7: 13–26.

———. 1989. Myxomycetes from Greenland. *Opera Botanica* 100: 93–103.

———. 1990. Notes on Icelandic myxomycetes. *Acta Botanica Islandica* 10: 3–21.

Gray, W. D., and C. J. Alexopoulos. 1968. *Biology of the Myxomycetes*. The Ronald Press Company, New York.

Greenslade, P., S. Moore, and R. Farrow. 2002. Observations on the feeding behavior of Uchidanurinae (Collembola: Neauridae) in Australia. *The Victorian Naturalist* 119: 221–223.

Hagelstein, R. 1936. A critical study of the mycetozoan of Long Island. *Mycologia* 28: 547–622.

———. 1944. *The Mycetozoa of North America Based Upon the Specimens in the Herbarium of the New York Botanical Garden*. Published by the author, Mineola, New York.

Härkönen, M. 1979. A check-list of Finnish myxomycetes. *Karstenia* 19: 8–18.

———. 1981a. Ten myxomycetes species new to Finland. *Karstenia* 21: 53–56.

———. 1981b. Myxomycetes developed on litter of common Finnish trees in moist chamber cultures. *Nordic Journal of Botany* 1: 791–794.

———. 1989. Thirteen taxa of myxomycetes new to Finland. *Karstenia* 28: 93–99.

Härkönen, M., and T. Saarimäki. 1991. Tanzanian myxomycetes: first survey. *Karstenia* 31: 31–54.

Härkönen, M., and T. Ukkola. 2000. Conclusions on myxomycetes compiled over twenty-five years from 4793 moist chamber cultures. *Stapfia* 155: 105–112.

Horak, E. 1966. Sobre dos nuevos especies de hongos recolectados en el Antarctico. *Contribucion del Antartica Argentinto* 104: 3–13.

Howard, F. L., and M. E. Currie. 1932a. Parasitism of myxomycete plasmodia on the sporophores of hymenomycetes. *Journal of the Arnold Arboretum* 13: 270–284.

———. 1932b. Parasitism of myxomycete plasmodia on fungus mycelia. *Journal of the Arnold Arboretum* 13: 438–446.

Hudson, H. J. 1986. *Fungal Biology*. Edward Arnold, Baltimore, Maryland.

Illana, C., M. Heykoop, and G. Moreno. 1990. Contribution to the study of the myxomycetes in Spain. III. Catalogue of myxomycetes of Spain. *Mycotaxon* 38: 37–69.

Indira, P. U. 1968. Some slime moulds from southern India IX. *Journal of the Indian Botanical Society* 47: 330–341.

Ing, B. 1974. Mouldy myxomycetes. *Bulletin of the British Mycological Society* 8: 25–30.

———. 1976. More on mouldy myxomycetes. *Bulletin of the British Mycological Society* 10: 30.

———. 1980. A revised census catalogue of British myxomycetes, Part 1. *Bulletin of the British Mycological Society* 14: 97–111.

———. 1982. A revised census catalogue of British myxomycetes, Part 2. *Bulletin of the British Mycological Society* 16: 26–35.

———. 1999. *The Myxomycetes of Britain and Ireland: an Identification Handbook*. The Richmond Publishing Company, Ltd., Slough, England.

Ing, B., and R. I. L. Smith. 1980. Two myxomycetes from South Georgia. *British Antarctic Survey Bulletin* 50: 118–120.

———. 1983. Further myxomycete records from South Georgia and the Antarctic Peninsula. *British Antarctic Survey Bulletin* 59: 80–81.

Johannesen, E. W. 1984. New and interesting myxomycetes from Norway. *Nordic Journal of Botany* 4: 513–520.

Johnson, L. M. 2021. What slime knows. *Orion magazine*, August 2021.

Jordan, C. C., M. H. Brims, E. J. Speijers, and E. M. Davison. 2006. Myxomycetes on the bark of *Banksia attenuata* and *B. menziesii* (Proteaceae). *Australian Journal of Botany* 54: 357–365.

Keller, H. W. 1973. Myxomycetes from the Everglades National Park and adjacent areas. *Ohio Journal of Science* 73: 364–369.

Keller, H. W., and K. L. Braun. 1977. Myxomycetes of Mexico II. *Boletin de la Sociedad Mexicana de Micrologia* 11: 167–180.

Keller, H. W., and T. E. Brooks. 1973. Corticolous myxomycetes I. Two new species of *Didymium*. *Mycologia* 65: 286–294.

———. 1977. Corticolous myxomycetes VII: contributions toward a monograph of *Licea*, five new species. *Mycologia* 69: 667–684.

Keller, H. W., and U. H. Eliasson. 1992. Taxonomic evaluation of *Perichaena depressa* and *P. quadrata* based on controlled cultivation, with additional observations on the genus. *Mycological Research* 96: 1085–1097.

Keller, H. W., S. E. Everhart, and C. M. Kilgore. 2021. The myxomycetes: introduction, basic biology, life cycles, genetics, and reproduction. Pages 1–45 in Rojas, C., and S. L. Stephenson (eds.) *Myxomycetes. Biology, Systematics, Biogeography and Ecology*, 2nd ed. Elsevier Academic Press, Cambridge, Massachusetts.

Keller, H. W., and K. L. Snell. 2002. Feeding activities of slugs on myxomycetes and macrofungi. *Mycologia* 94: 757–760.

Knight, K. J., and M. H. Brims. 2010. Myxomycota census of Western Australia. *Nuytsia* 20: 283–307.

Ko Ko, T. W., W. C. Rosing, Z. Z. W. Ko Ko, and S. L. Stephenson. 2013. Myxomycetes of Myanmar. *Sydowia* 65: 267–276.

Ko Ko, T. W., H. T. M. Tran, M. E. Clayton, and S. L. Stephenson. 2012. First records of myxomycetes in Laos. *Nova Hedwigia* 96: 73–81.

Ko Ko, T. W., H. T. M. Tran, S. L. Stephenson, D. W. Mitchell, C. Rojas, K. D. Hyde, and S. Lumyong. 2010. Myxomycetes of Thailand. *Sydowia* 62: 243–260.

Kowalski. D. T. 1967. Observations on the Dianemaceae. *Mycologia* 59: 1075–1084.

———. 1970. The species of *Lamproderma*. *Mycologia* 62: 621–672.

———. 1971. The genus *Lepidoderma*. *Mycologia* 63: 490–516.

Kryvomaz, T., A. Michaud, and S. L. Stephenson. 2020. An annotated checklist of myxomycetes from the Seychelles Islands, Indian Ocean. *Karstenia* 58: 215–240.

Kryvomaz, T., H. T. M. Tran, and C. Rojas. 2021. Uses, applications, and disciplinary integrations using myxomycetes. 2021. Molecular techniques and current research approaches. Pages 493–531 in Rojas, C., and S. L. Stephenson, eds. *Myxomycetes. Biology, Systematics, Biogeography and Ecology*, 2nd ed. Elsevier Academic Press, Cambridge, Massachusetts.

Kylin, H. 2001. Biodiversity in a slime mould: arthropods associated with *Brefeldia maxima*. *Mycologist* 15: 70–73.

Lado, C. 1994. A checklist of myxomycetes of the Mediterranean countries. *Mycotaxon* 52: 117–185.

———. 2005–2022. *An online nomenclatural information system of Eumycetozoa*. http://www.nomen.eumycetozoa.com (15 June 2022).

Lado, C., A. Estrada-Torres, S. L. Stephenson, D. Wrigley de Basanta, and M. Schnittler. 2003. Biodiversity assessment of myxomycetes from two tropical forest reserves in Mexico. *Fungal Diversity* 12: 67–110.

Lado, C., J. Mosquera, A. Estrada-Torres, E. Beltrán-Tejera, and D. Wrigley de Basanta. 2007. Description and culture of a new succulenticolous *Didymium* (Myxomycetes). *Mycologia* 99: 602–611.

Lado, C., and F. Pando. 1997. Myxomycetes, Ceratiomyxales, Echinosteliales, Liceales, Trichiales. *Flora Mycologica Iberica* 2. 1–233.

Lado, C., and D. Wrigley de Basanta. 2008. A review of Neotropical Myxomycetes. *Anales del Jardín Botánico de Madrid* 65: 211–254.

Lado, C., D. Wrigley de Basanta, and A. Estrada-Torres. 2011. Biodiversity of Myxomycetes from the Monte desert of Argentina. *Anales del Jardín Botánico de Madrid* 68: 61–95.

Lado, C., D. Wrigley de Basanta, A. Estrada-Torres, and S. L. Stephenson. 2016. The myxomycete diversity of the coastal deserts of Peru with emphasis on the lomas formations. *Anales del Jardín Botánico de Madrid* 73(1): e032 2016. doi: http://dx.doi.org/10.3989/ajbm.2436

———. 2013. The biodiversity of Myxomycetes in Central Chile. *Fungal Diversity* 59: 3–32.

Lakhanpal, T. N., and K. G. Mukerji. 1981. Taxonomy of the Indian Myxomycetes. *Bibliotheca Mycologia.* Vaduz J. Cramer. FL-9490. Germany.

Largent, D., D. Johnson, and R. Watling. 1977. *How to Identify Mushrooms to Genus III: Microscopic Features.* Mad River Press, Inc., Eureka, California.

Lawrence, J. F., and A. F. Newton, Jr. 1980. Coleoptera associated with the fruiting bodies of slime molds (Myxomycetes). *Coleopterists Bulletin* 34: 129–143.

Lazo, W. R. 1961. Growth of green algae with myxomycete plasmodia. *American Midland Naturalist* 65: 381–383.

———. 1966. Notes and illustrations of myxomycetes from Chile and other countries. *Mycologia* 58: 67–79.

Leontyev, D.V., M. Buttgereit, A. Kochergina, A., G. Shchepin, and M. Schnittler. 2022a. Two independent genetic markers support separation of the myxomycete *Lycogala epidendrum* into numerous biological species. *Mycologia* http://doi.org/10.1080/00275514.2022.2133526

Leontyev, D.V., Y. Ishchenko, and M. Schnittler. 2023. Fifteen new species from the myxomycete genus *Lycogala. Mycologia* 115: 524–560.

Leontyev, D. V., and M. Schnittler. 2017. The phylogeny of myxomycetes. Pages 83–105 in Stephenson, S. L., and C. Rojas, eds. Myxomycetes. *Biology, Systematics, Biogeography and Ecology*. Elsevier Academic Press, Cambridge, Massachusetts.

Leontyev, D. V., M. Schnittler, and S. L. Stephenson. 2014a. Pseudocapillitium or true capillitium? Assessment of capillitial structures in *Alwisia bombarda. Nova Hedwigia* 99: 441–451.

———. 2015. A critical revision of the *Tubifera ferruginosa* complex. *Mycologia* 107: 959–985.

Leontyev, D. V., M. Schnittler.,Y. Ishchenko, A. Quade, H. Kahlert, C. R. Alvarado, and S. L. Stephenson. 2022b. Another species complex in myxomycetes: diversity of peridial structures in *Lycogala epidendrum. Nova Hedwigia* 114: 413–434.

Leontyev, D. V., M. Schnittler, S. L. Stephenson, Y. K. Novozhilov, and O. N. Shchepin. 2019. Towards a phylogenetic classification of the myxomycetes. *Phytotaxa* 399: 209–238.

Leontyev, D. V., S. L. Stephenson, and M. Schnittler. 2014b. A new species of *Alwisia* (Myxomycetes) from New South Wales and Tasmania. *Fungal Diversity* 106: 1212–1219.

Li, T., and H.Z. Li. 1989. Myxomycetes from China. I. *Mycotaxon* 35: 429–436.

Lindley, L. A., S. L. Stephenson, S. L., and F. W. Spiegel. 2007. Protostelids and myxomycetes isolated from aquatic habitats. *Mycologia* 99: 504–509.

Link, J. H. F. 1833. Handbuch zur erkennung der nutzbarsten und am häufigsten vorkommenden Gewäches Vol. 3. Ordo Fungi, Subordo 6. *Myxomycetes* 405–422, 432–433. Berlin, Germany.

Lister, A. 1888. Notes on the plasmodium of *Badhamia utricularis* and *Brefeldia maxima. Annals of Botany* 2: 124.

———. 1894. *A Monograph of the Mycetozoa.* British Museum of Natural History, London.

———. 1911. *A Monograph of the Mycetozoa*. 2nd ed. (revised by G. Lister). British Museum of Natural History, London.

———. 1925. *A Monograph of the Mycetozoa.* 3rd ed. (revised by G. Lister). British Museum of Natural History, London.

Lister, G. 1918. The Mycetozoa: a short history of their study in Britain; an account of their habits generally; and a list of species recorded from Essex. *Essex Field Club Special Memoirs* 6: 1–54.

Lister, G. 1938. The Rev. William Cran and his scientific work. *Journal of Botany* 76: 319–327.

López, A., H. Garcia, and J. L. Herrador. 1982. Nuevos registros de hongos comestibles de la region del cofre de perote, Estado de Veracruz. In *Primer Congreso Nacional de Micologia*. Veracruz, Mexico: *Sociedad Mexicana de Micologia*.

Lowy, B. 1953. Myxomycetes of Louisiana. *Mycologia* 45: 926–933.

Lloyd, S. J. 2014. *Where the Slime Mould Creeps: the Fascinating World of Myxomycetes*. Tympanocryptis Press, Tasmania, Australia.

———. 2018. *Where the Slime Mould Creeps: the Fascinating World of Myxomycetes*, 2nd ed. Tympanocryptis Press, Tasmania, Australia

Macabago, S. A. B., T. E. E. Dela Cruz, and S. L. Stephenson. 2012. First records of myxomycetes from Lubang Island, Occidental Mindoro, Philippines. *Sydowia* 64: 109–118.

Macbride, T. H. 1899. *North American Slime-Moulds*. Macmillan Company, New York.

Macbride, T. H., and G. W. Martin. 1934. *The Myxomycetes*. Macmillan Company, New York.

Madelin, M. F. 1984. Myxomycetes, microorganisms and animals: a model of diversity in animal-microbial interactions. Pages 1–33 in *Invertebrate-Microbial Interactions*. Anderson, J. N., A. D. A. Rayner, and D. W. H. Walton, eds. Cambridge University Press, New York.

Maimoni-Rodella, R. C. S., and G. Gottsberger. 1980. Myxomycetes from the forest and the cerrado vegetation in Botucatu, Brazil: a comparative ecological study. *Nova Hedwigia* 34: 207–246.

Martin, G. W. 1949. Class Myxomycetes. *North American Flora* 1 (part 1). New York Botanical Garden, New York, New York.

Martin, G. W., and C. J. Alexopoulos. 1969. *The Myxomycetes*. University of Iowa Press, Iowa City.

Martin, G. W., C. J. Alexopoulos, and M. L. Farr. 1983. *The Genera of Myxomycetes*. Iowa City: University of Iowa Press.

McManus, M. A. 1962. Some observations on plasmodia of the Trichiales. *Mycologia* 54: 78–90.

Mele, A. 2020. *A Companion to Cosmopolitan Slime Molds*. Published by the author.

Merrill, R. A. 1969. *A Preliminary Survey of the Coprophilous Myxomycetes of California*. M.S. thesis, Chico State College, Chico, California.

Mitchel, D. H., S. W. Chapman, and M. L. Farr. 1980. Notes on Colorado fungi IV: myxomycetes. *Mycotaxon* 10: 299–349.

Mitchell, D.W. 1995. The Myxomycota of Australia. *Nova Hedwigia* 60: 269–295.

———. 1980. *A Key to the Corticolous Myxomycetes*. Cambridge, England: The British Mycological Society.

———. 1992. The Myxomycota of New Zealand and its island territories. *Nova Hedwigia* 55: 231–256.

Morris, E. F. 1954. Notes on Wyoming myxomycetes. *University of Wyoming Publications* 18: 83–90

Nannenga-Bremekamp, N. E. 1991. *A Guide to Temperate Myxomycetes*. (Translated by Burggraaf, Y., and A. Feest.) Biopress Limited, England.

Ndiritu, G. G., K. E. Winsett, F. W. Spiegel, and S. L. Stephenson. 2009a. A checklist of African myxomycetes. *Mycotaxon* 107: 353–356.

Ndiritu, G. G., F. W. Spiegel, and S. L. Stephenson. 2009b. Distribution and ecology of the assemblages of myxomycetes associated with major vegetation types in Big Bend National Park, USA. *Fungal Ecology* 2: 168–183.

Neubert, H., W. Nowotny, and K. Baumann. 1993. *Die Myxomyceten*. Bd. 1. Gomaringen, Baumann.

Neubert, H., Nowotny, W., Baumann, K., Marx, H. 1995. *Die Myxomyceten*. Bd. 2. Gomaringen, Baumann.

———. 2000. *Die Myxomyceten*. Bd. 3. Gomaringen, Baumann.

Nguyen, L. T. T., O. Sanchez-Mahecha, K. J. Almadrones-Reyes, J. C. Redeña-Santos, and N. H. A. Dagamac. 2019. Occurrence of leaf litter inhabiting myxomycetes from lowland forest patches of northern and central Vietnam. *Tropical Ecology* 60: 495–506.

Nikolov, B. P., and I. P. Hristova. 2009. Food storage of slime mould in Siberian Jay *Perisoreus infaustus* during the post-breeding season. *Ornis Secica* 19: 59–91.

Novozhilov, Y. K., A. W. Rollins, and M. Schnittler. 2017. Ecology and Distribution of Myxomycetes. Pages 253–297 in Stephenson, S.L. and C. Rojas, eds. *Myxomycetes: Biology, Systematics, Biogeography and Ecology*. Elsevier Academic Press, Cambridge, Massachusetts.

Novozhilov, Y. K., A. W. Rollins, O. N. Schepin, and M. Schnittler. 2021. Ecology and distribution of myxomycetes. Pages 325–376 in Rojas, C., and S. L. Stephenson, eds. *Myxomycetes. Biology, Systematics, Biogeography and Ecology*, 2nd ed. Elsevier Academic Press, Cambridge, Massachusetts.

Novozhilov, Y. K., D. A. Erastova, O. N. Shchepin, M. Schnittler, A. V. Aleksandrova, E. S. Popov, and A. N. Kuznetsov. 2017. Myxomycetes associated with monsoon lowland tall tropical forests in southern Vietnam. *Nova Hedwigia* 104: 143–182.

Novozhilov, Y. K., D. W. Mitchell, and M. Schnittler. 2003. Myxomycete biodiversity of the Colorado Plateau. *Mycological Progress* 2: 243–258.

Novozhilov, Y. K., M. Schnittler, A. W. Rollins, and S. L. Stephenson. 2000. Myxomycetes from different forest types in Puerto Rico. *Mycotaxon* 77: 285–299.

Olive, L. S. 1975. *The Mycetozoans*. Academic Press, New York.

Onduso, F. N., S. Tewari, L. Tewari, and S. L. Stephenson. 2019. Myxomycete diversity on bison dung and cow dung in the Missouri River watershed of the Standing Rock Sioux Reservation, North Dakota. *Journal of Bacteriology and Mycology* 7: 75–79.

Parker, H. 1946. Studies in the nutrition of some aquatic myxomycetes. *Journal of the Elisah Mitchell Scientific Society* 62: 231–247.

Pelloquin, R. A., and S. L. Stephenson. 2016. Assessing the ecological impact of whole watershed acidification on the myxomycetes associated with forest floor leaf litter. *Current Research in Environmental & Applied Mycology* 6: 184–189.

Poulain, M., M. Meyer, and J. Bozonnet. 2011. *Les Myxomycètes*. Fédération Mycologique et Botanique. Dauphiné-Savoie, Delémont.

Pratt, R. 1934. Fruiting relations of some mycetozoa. *Torreya* 34: 37–39.

Ranade, V. D., S. T. Korade, A. V. Jagtap, and K. R. Ranadive. 2012. Checklist of Myxomycetes from India. *Mycosphere* 3: 358–390.

Raper, K. B. 1984. *The Dictyostelids*. Princeton University Press, Princeton, New Jersey.

Richardson, K. A., and R. S. Currah. 1990. Myxomycetes of Alberta. *Mycotaxon* 37: 363–378.

Robbrecht, E. 1974. The genus *Arcyria* Wiggers (Myxomycetes) in Belgium. *Bulletin du Jardin Botanique Belgique* 44: 303–353.

Rogerson, C. T., and S. L. Stephenson. 1993. Myxomyceticolous fungi. *Mycologia* 85: 456–469.

Rojas, C., C. Lado, and P. A. Rojas. 2018. Myxomycete diversity in Costa Rica. *Mycosphere* 9: 227– 255.

Rojas, C., P. A. Rojas, and S. L. Stephenson. 2021. Phenology of myxomycetes in Turrialba, Costa Rica. *Karstenia* 59(1-2): 1–12.

Rojas, C., and S. L. Stephenson. 2008. Myxomycete ecology along an elevation gradient on Cocos Island, Costa Rica. *Fungal Diversity* 29: 117-127.

Rojas, C., and S. L. Stephenson, eds. 2021. *Myxomycetes. Biology, Systematics, Biogeography and Ecology*, 2nd ed. Elsevier Academic Press, Cambridge, Massachusetts.

Rojas, C., S. L. Stephenson, and P. A. Rojas. 2020. Myxomycete colonization on translocated and nontranslocated dead leaves from temperate and tpical forests. *Studies in Fungi* 5: 462-470.

Rollins, A. W., and S. L. Stephenson. 2012. Myxogastrid distribution within the leaf litter microhabitat. *Mycosphere* 3: 543–549.

Rollins, A. W., and S. L. Stephenson. 2013. Myxomycetes associated with grasslands of the western central United States. *Fungal Diversity* 59: 147–158.

Rostafinski, J. T. 1874–1876. Sluzowce (Mycetozoa) Monografia. *Towarzystwo Nauk Ścisłych w Paryżu* 5: 1–215 (1974); 217–432 (1895); Dodatek [appendix] 8: 1–43 (1876).

Russell, D. R., and M. B. Huneycott. 1972. A checklist of the myxomycetes of Mississippi. *Journal of Mississippi Academy of Science* 17: 14.

Russell, L. K. 1979. Beetles associated with slime molds (Mycetozoa) in Oregon and California (Coleoptera: Leiodidae, Sphindidae, Lathridiidae). *Pan-Pacific Entomologist* 55: 19.

Saccardo, P. A., C. H. Peck, and W. Trelease. 1904. The fungi of Alaska. In Harriman Alaska Expedition Series, vol. 5. *Cryptogamic Botany*. Doubleday, Page and Company. New York: 1364.

Samuels, G. J. 1973. The myxomyceticolous species of Nectria. *Mycologia* 65: 401–420.

———. 1988. Fungicolous, lichenicolous and myxomyceticolous species of *Hypocreopsis, Nectriopsis, Nectria, Peristomialis,* and *Trichonectria. Memoirs of the New York Botanical Garden* 48: 178.

Santesson, R. 1964. Swedish myxomycetes. *Svensk Botanisk Tidskrift* 58: 113–124.

Schnittler, M. 2001a. Ecology of myxomycetes from a winter-cold desert in western Kazakhstan. *Mycologia* 93: 135–167.

Schnittler, M. 2001b. Foliicolous liverworts as a microhabitat for Neotropical myxomycetes. *Nova Hedewigia* 72: 259–270.

Schnittler, M., and S. L. Stephenson. 2002. Inflorescences of Neotropical herbs as a newly discovered microhabitat for myxomycetes. *Mycologia* 94: 6–20.

Schnittler, M., C. Lado, and S. L. Stephenson. 2002. Rapid biodiversity assessment of a tropical myxomycete assemblage—Maquipucuna Cloud Forest Reserve, Ecuador. *Fungal Diversity* 9: 135–167.

Schnittler, M., M. Unterseher, T. Pfeiffer, Y. K. Novozholov, and A. M. Fiore-Donno. 2010. Ecology of sandstone ravine myxomycetes from Saxonian Switzerland (Germany). *Nova Hedwigia* 90: 277–302

Schoknecht, J. D., and E. B. Small. 1972. Scanning electron microscopy of the acellular slime molds (Mycetozoa-Myxomycetes) and the taxonomic significance of surface morphology of spores and accessory structures. *Transactions of the American Microscopy Society* 91: 380–410.

Sellier, R., and M. Chassain. 1976. Observations sur le mode de dissémination des spores d'un champignon myxomycète par un insecte diptère mycètophilidae. *Bulletin de la Société Sciences Naturelle de l'Ouest de France* 74: 81–85.

Sharnoff, S. D. 1991. Beauties from a beast: woodland Jekyll and Hydes. *Smithsonian* 22: 98–103.

Shearer, C. A., and J. L. Crane. 1986. Illinois fungi XII. Fungi and myxomycetes from wood and leaves submerged in southern Illinois swamps. *Mycotaxon* 25: 527–538.

Shrikh, S., M. Thulin, J. C. Cavender, R. E. Hernández, S.-I. Kawakami, C. Lado, J. C. Landolt, V. Nanjundiah, D. Queller, P. Shaap, J. Strassmann, F. Spiegel, S. L. Stephenson, E. M. Vadell, and S. L Baldauf. 2018. A new classification of the dictyostelids. *Protist* 169: 1–28.

Smith, E. C. 1929. The longevity of myxomycete spores. *Mycologia* 21: 321–323.

Smith, H. V., A. H. Smith, and N. S. Weber. 1981. *How to Know the Non-Gilled Mushrooms*. Wm. C. Brown Company, Dubuque, Iowa.

Smith, T., and S. L. Stephenson. 2007. Algae associated with myxomycetes and leafy liverworts on decaying spruce logs. *Castanea* 72: 50–57.

Stephenson, S. L. 1982. West Virginia myxomycetes. *Mycotaxon* 15: 455–458.

———. 1983. Myxomycetes associated with southern Appalachian spruce-fir forests. *Proceedings of the West Virginia Academy of Science* 55: 118–124.

———. 1985. Myxomycetes in the laboratory II: moist chamber cultures. *American Biology Teacher* 47: 487–489.

———. 1988. Distribution and ecology of myxomycetes in temperate forests. I. Patterns of occurrence in the upland forests of south-western Virginia. *Canadian Journal of Botany* 66: 2187–2207.

———. 1989. Distribution and ecology of myxomycetes in temperate forests. II. Patterns of occurrence on bark surface of living trees, leaf litter, and dung. *Mycologia* 81: 608–621.

———. 2003. *Myxomycetes of New Zealand*. Fungal Diversity Press, Hong Kong.

———. 2004. Distribution and ecology of myxomycetes in southern Appalachian subalpine coniferous forests. *Memoirs of the New York Botanical Garden* 89: 203–212.

———. 2009. First records of myxomycetes from Ascension Island. *Sydowia* 61: 105–115.

———. 2021a. Myxomycetes recorded from the vicinity of the Mountain Lake Biological Station, southwestern Virginia. *Castanea* 86: 28–36.

———. 2021b. *Secretive Slime Moulds: The Myxomycetes of Australia*. CSIRO Publishing, Canberra.

Stephenson, S. L., and A. Feest. 2012. Ecology of soil Eumycetozoans. *Acta Protozoologica* 51: 201–208.

Stephenson, S. L., Fiore-Donno, A.-M., Schnittler, M. 2011. Myxomycetes in soil. *Soil Biology and Biochemistry* 43: 2237–2242.

Stephenson, S. L., G. Kaur, N. Payal, T. F. Elliott, and K. Vernes. 2020. Myxomycetes associated with arid habitats in northeastern South Australia. *Transactions of the Royal Society of South Australia* 144: 139–153.

Stephenson, S. L., and J. C. Landolt. 1998. Dictyostelid cellular slime molds in canopy soils of tropical forests. *Biotropica* 30: 657–661.

———. 2011. Dictyostelids from aerial "canopy soil" microhabitats. *Fungal Ecology* 4: 191–195.

———. 2015. Occurrence of myxomycetes in the aerial "canopy soil" microhabitat. *Mycosphere* 6: 83–86.

Stephenson, S. L., J. C. Landolt, and D. L. Moore. 1999. Protostelids, dictyostelids, and myxomycetes in the litter microhabitat of the Experimental Forest, Puerto Rico. *Mycological Research* 103: 209–214.

Stephenson, S. L., and G. A. Laursen. 1993. A preliminary report on the distribution and ecology of myxomycetes in Alaskan tundra. *Bibliotheca Mycologia* 150: 251–257.

Stephenson, S. L., G. A. Laursen, and R. D. Seppelt. 2007. Myxomycetes of subantarctic Macquarie Island. *Australian Journal of Botany* 55: 439–449.

Stephenson, S. L., and Y. K. Novozhilov. 2021. Coprophilous myxomycetes associated with herbivore dung in the northern Rocky Mountains. *Sydowia* 74: 175–180.

Stephenson, S. L., Y. K. Novozhilov, and M. Schnittler. 2000. Distribution and ecology of myxomycetes in high-latitude regions of the northern hemisphere. *Journal of Biogeography* 27: 741–754.

Stephenson, S. L., and C. Rojas. 2017. *Myxomycetes: Biology, Systematics, Biogeography, and Ecology*. Elsevier Academic Press, Cambridge, Massachusetts.

———. 2020. Mosses as spore traps for myxomycetes. *Sydowia* 72: 215–219.

Stephenson, S. L., M. Schnittler, D. W. Mitchell, and Y. K. Novozhilov. 2001. Myxomycetes of the Great Smoky Mountains National Park. *Mycotaxon* 78: 1–15.

Stephenson, S. L., R. D. Seppelt, and G. A. Laursen. 1992. The first record of a myxomycete from subantarctic Macquarie Island. *Antarctic Science* 4: 431–432.

Stephenson, S. L., and J. D. L. Shadwick. 2009. Nivicolous myxomycetes from alpine areas of south-eastern Australia. *Australian Journal of Botany* 57: 116–122.

Stephenson, S. L., R. W. Stauffacher, and C. Rojas. 2020. Myxomycetes collected in the eastern United States and the relative abundance of particular species. *Mycology* https://doi:org/10.1080/21501203.2019.1710392

Stephenson, S. L., and H. Stempen. 1994. *Myxomycetes: A Handbook of Slime Molds*. Timber Press, Portland, Oregon.

Stephenson, S. L., and B. C. Stephenson. 2019. Distribution and ecology of myxomycetes on Christmas Island, Indian Ocean. *Phytotaxa* 416: 138-148.

———. 2020. Biodiversity of myxomycetes on Norfolk Island, South Pacific Ocean. *Cunninghamia* 20: 115–121.

———. 2022a. Myxomycetes associated with the aerial litter microhabitat in a temperate deciduous forest. *Slime Molds* V2A6: 1–6.

———. 2022b. Myxomycetes associated with bryophyte mats in the Mountain Lake area of southwestern Virginia. *Slime Molds* V2A12: 1–5.

Stephenson, S. L., and S. M. Studlar. 1985. Myxomycetes fruiting upon bryophytes: coincidence or preference? *Journal of Bryology* 13: 537–548.

Stephenson, S. L., Q. D. Wheeler, J. V. McHugh, and P. R. Fraissinet. 1994. New North American associations of Coleoptera with myxomycetes. *Journal of Natural History* 28: 921–936.

Stevens, R., ed. 1974. *Mycology Guidebook*. Seattle: University of Washington Press.

Sugiura, S., Y. Fukasawa, R. Ogawa, S.-I. Kawakami, and Y. K. Yamazaki. 2019. Beetles as potential dispersers of slime mold spores. *Bulletin of the Ecological Society of America* 100(3). doi.org/10.1002/bes2.1564.

Sun, Y., C. Ma, and S. Halgamuge. 2017. The node-weighted steiner tree approach to identify elements of cancer-related signaling pathways. *BMC Bioinformatics* 18: 551. https://doi/org/10.1186/s12859-017-1958-4

Sutherland, J. B., and R. I. Crawford. 1979. Gray jay feeding on slime mold. *The Murrelet* 60: 28.

Tamayama, M. 2000. Nivicolous taxa of the myxomycetes in Japan. *Stapfia* 73: 121–129.

Thom, C., and K. B. Raper. 1930. Myxamoebae in soil and decomposing crop residues. *Journal of the Washington Academy of Science* 20: 362–370.

Thompson, A. R., G. S. Powell, and B. J. Adams. 2019. Provisional checklist of terrestrial heterotrophic protists from Antarctica. *Antarctic Science* 31: 287–303.

Tran, H. T. M., S. L. Stephenson, K. D. Hyde, and O. Mongkolporn. 2008. Distribution and occurrence of myxomycetes on agricultural ground litter and forest floor litter in Thailand. *Mycologia* 100: 181–190.

Ukkola, T., M. Härkönen, and T. Saarimaki. 1996. Tanzanian Myxomycetes: second survey. *Karstenia* 36: 51–77.

Uribe-Meléndez, J. 1995. Catálogo de los Myxomycetes registrados para Colombia. *Caldasia* 18: 23–36.

Urich, T., A. Lanzén, J. Qi, D. H. Huson, C. Schleper, and S. C. Schuster. 2008. Simultaneous assessment of soil microbial community structure and function through analysis of the meta-transcriptome. *PLoS ONE* 3 doi.org/10.1371/journal.pone.0002527.

Villarreal, L. 1983. Algunas especies de myxomycetes no registradas del Estado de Veracruz. *Boletín de la Sociedad Méxicana de Micologia* 18: 153-164.

Walker, L. M., T. Hoppe, and M. L. Silliker. 2021. Molecular techniques and current research approaches. Pages 195–229 *in* Rojas, C., and S. L. Stephenson (eds.) *Myxomycetes. Biology, Systematics, Biogeography and Ecology*, 2nd ed. Elsevier Academic Press, Cambridge, Massachusetts.

Walker, L. W., and S. L. Stephenson. 2016. The species problem in the myxomycetes revisited. *Protist* 167: 319–338.

Warcup, J. H. 1950. The soil-plate method for isolation of fungi from soil. *Nature* 166: 117–118.

Ward, H. M. 1886. The morphology and physiology of an aquatic myxomycete. *Studies from the Biological Laboratories of Owens College* 1: 64–86.

Weaver, R. F. 1976. The cancer puzzle. *National Geographic* 150: 396–399.

Welden, A. L. 1951. A taxonomic survey of myxomycetes of the Great Smoky Mountains National Park. *Journal of the Tennessee Academy of Science* 26: 271–276.

Wellman, P. 2016. Distinctive corticolous myxomycete assemblages of tropical, temperate and arid regions of Australia. *Cunninghamia* 16: 101–114.

———. 2017. Temperate-climate myxomycetes in Australia and the Southern Hemisphere. *Cunninghamia* 17: 1–13.

Wheeler, Q. D. 1979. Slime mold beetles of the genus *Anisotoma* (Leiodidae): classification and evolution. *Systematic Entomology* 4: 251–309.

———. 1984a. Evolution of slime mold feeding in leiodid beetles. Pages 446-477 in *Fungus-Insect Relationships: Perspectives in Ecology and Evolution*, eds. Wheeler, Q. D., and M. Blackwell. Columbia University Press, New York.

———. 1987. A new species of *Agathidium* associated with an "epimycetic" slime mold plasmodium on *Pleurotus* fungi (Coleoptera: Leiodidae—Myxomycetes: PhysaralesBasidiomycetes: Tricholomataceae). *Coleopterists Bulletin* 41: 395–404.

———. 1984b. Associations of beetles with slime molds: ecological patterns in the Anisotomini (Leiodidae). *Bulletin of the Entomological Society of America* 30: 14–18.

Whitney, K. D. 1980. The myxomycete genus *Echinostelium*. *Mycologia* 72: 950–987.

———. 1982. A survey of the corticolous myxomycetes of California. *Madroño* 29: 259–268.

Winsett, K. E., and S. L. Stephenson. 2013. Myxomycetes isolated from submerged plant material collected in the Big Thicket National Preserve, Texas. *Mycosphere* 4: 227–231.

Wrigley de Basanta, D., and C. Lado. 2005. A taxonomic evaluation of the stipitate *Licea* species. *Fungal Diversity* 20: 261–314.

Wrigley de Basanta, D., C. Lado, A. Estrada-Torres, and S. L. Stephenson. 2010. Biodiversity of myxomycetes in subantarctic forests of Patagonia and Tierra del Fuego, Argentina. *Nova Hedwigia* 90: 45–79.

Wrigley de Basanta, D., S. L. Stephenson, C. Lado, A. Estrada-Torres, and A. M. Nieves-Rivera. 2008. Lianas as a microhabitat for myxomycetes in tropical forests. *Fungal Diversity* 28: 109–125.

Yamamoto, Y. 2021. *Biota of Japanese Myxomycetes*. Committee for the publication of "Biota of Japanese Myxomycetes," Tsukuba-mirai City, Ibaraki, Japan.

Yamamoto, Y., and H. Hagiwara. 1990. Myxomycetes from central Nepal II. Pages 33–40 in *Cryptogams of the Himalayas, vol. 2, Central and Eastern Nepal*. M. Watanabe and S. B. Malla, eds. National Science Museum, Tsukuba, Japan.

Zenlyanskaya, I., Y. K. Novozhilov, and M. Schnittler. 2020. An annotated checklist slime molds (Myxomycetes = Myxogastrea) of western Kazakhstan. *Karstenia* 58: 168–189.

Index

Q-R

S

T

U-Z

About the Author

Steve Stephenson retired from the University of Arkansas, where he taught courses in plant biology, plant ecology, forest ecology, and mycology as well as serving as the mentor for twenty-two graduate students who received their M.S. or Ph.D. degrees. He received his B.S. in biology from the University of Lynchburg and both an M.S and a Ph.D. from Virginia Tech. He was on the faculty of Fairmont State College (now University) for 27 years before moving to the University of Arkansas in 2003. During his academic career, he was a Senior Fulbright Scholar at Himachal University in India in 1987, a Visiting Scientist at the Australian Antarctic Division in 1995, and the William Evans Visiting Fellow at the University of Otago (Dunedin, New Zealand) in 2003. His studies of myxomycetes (slime molds) have taken him to all seven continents and every major type of terrestrial ecosystem. He is the author or coauthor of 19 scientific books and more than 500 book chapters and papers in peer-reviewed journals. Among his books are *Myxomycetes: Biology, Systematics, Biogeography, and Ecology* (2017). *Mushrooms of the Southeastern United States* (2018), and *Secretive Slime Moulds: Myxomycetes of Australia* (2021). In addition to his scientific books, he has written more than 20 science fiction novels.

Made in the USA
Monee, IL
27 October 2026

66935975R00108